Huy The Nguyen

Nonlinear Behavior of Piezoceramics at Moderate Strains

Huy The Nguyen

Nonlinear Behavior of Piezoceramics at Moderate Strains

Quasi-static and Dynamic Behavior

Südwestdeutscher Verlag für Hochschulschriften

Impressum/Imprint (nur für Deutschland/only for Germany)
Bibliografische Information der Deutschen Nationalbibliothek: Die Deutsche Nationalbibliothek verzeichnet diese Publikation in der Deutschen Nationalbibliografie; detaillierte bibliografische Daten sind im Internet über http://dnb.d-nb.de abrufbar.
Alle in diesem Buch genannten Marken und Produktnamen unterliegen warenzeichen-, marken- oder patentrechtlichem Schutz bzw. sind Warenzeichen oder eingetragene Warenzeichen der jeweiligen Inhaber. Die Wiedergabe von Marken, Produktnamen, Gebrauchsnamen, Handelsnamen, Warenbezeichnungen u.s.w. in diesem Werk berechtigt auch ohne besondere Kennzeichnung nicht zu der Annahme, dass solche Namen im Sinne der Warenzeichen- und Markenschutzgesetzgebung als frei zu betrachten wären und daher von jedermann benutzt werden dürften.

Coverbild: www.ingimage.com

Verlag: Südwestdeutscher Verlag für Hochschulschriften GmbH & Co. KG
Heinrich-Böcking-Str. 6-8, 66121 Saarbrücken, Deutschland
Telefon +49 681 37 20 271-1, Telefax +49 681 37 20 271-0
Email: info@svh-verlag.de

Approved by: Berlin, TU Berlin, Dissertation, 2011

Herstellung in Deutschland:
Schaltungsdienst Lange o.H.G., Berlin
Books on Demand GmbH, Norderstedt
Reha GmbH, Saarbrücken
Amazon Distribution GmbH, Leipzig
ISBN: 978-3-8381-1333-3

Imprint (only for USA, GB)
Bibliographic information published by the Deutsche Nationalbibliothek: The Deutsche Nationalbibliothek lists this publication in the Deutsche Nationalbibliografie; detailed bibliographic data are available in the Internet at http://dnb.d-nb.de.
Any brand names and product names mentioned in this book are subject to trademark, brand or patent protection and are trademarks or registered trademarks of their respective holders. The use of brand names, product names, common names, trade names, product descriptions etc. even without a particular marking in this works is in no way to be construed to mean that such names may be regarded as unrestricted in respect of trademark and brand protection legislation and could thus be used by anyone.

Cover image: www.ingimage.com

Publisher: Südwestdeutscher Verlag für Hochschulschriften GmbH & Co. KG
Heinrich-Böcking-Str. 6-8, 66121 Saarbrücken, Germany
Phone +49 681 37 20 271-1, Fax +49 681 37 20 271-0
Email: info@svh-verlag.de

Printed in the U.S.A.
Printed in the U.K. by (see last page)
ISBN: 978-3-8381-1333-3

Copyright © 2011 by the author and Südwestdeutscher Verlag für Hochschulschriften GmbH & Co. KG and licensors
All rights reserved. Saarbrücken 2011

Abstract

Piezoceramic sensors and actuators have found broad fields of applications in recent decades. In the range of small strains resulting from weak electrical and/or mechanical loads, the behavior of piezoceramics is usually described by linear constitutive equations. Nonlinear hysteretic models are used to describe polarization processes or the behavior of piezoceramics in presence of strong electric fields and/or mechanical stresses above coercive magnitude giving rise to polarization switching processes.

On the other hand, nonlinear behavior of a softening Duffing-oscillator including jump phenomena or multiple stable amplitude responses at the same excitation voltage amplitude and frequency can be observed, when polarized piezoceramics are excited by weak electric fields far away from coercive ones. These are referred to as dynamic tests in the following. The present work is focused on the description of the nonlinear effects at ranges of moderate strains, as they occur typically in such dynamic tests. These nonlinear effects can classically be described by introducing nonconservative and higher-order terms into electric enthalpy or constitutive equations. Using the amplitude–frequency responses from dynamic experiments near resonance, the parameters of piezoceramics can be determined. Unfortunately, it is difficult to decide on some of the nonlinear characteristics, for example the type of conservative (mechanical, coupling or dielectric) nonlinearities or of nonconservative (damping) ones.

To overcome these problems, quasi-static experiments with moderate applied electric fields as well as tension and compression tests at moderate stresses resulting in strains of the same order as those in the dynamic cases are performed. Transversally polarized piezoceramics subjected to moderate quasi-static electric field in the polarization direction exhibit nonlinear hysteretic relations between the longitudinal strain or the electric displacement density and the applied fields. Stress–strain hysteretic behavior are also observed in tension and compression tests. These quasi-static responses can then be described by four of the most common hysteresis models, namely the classical Preisach model, the Prandtl-Ishlinskii model, the Masing model and the Bouc-Wen model, which are related to one another.

The Masing and Bouc-Wen models have the advantage to be described by the differential evolution equations of internal variables, so that these hysteresis models are easily integrated into the linear conservative modeling of longitudinal vibrations of piezoceramics. Finally, the mechanical nonlinearities can be determined directly from the results of tension and compression tests on the condition that the electrodes of piezoceramics are short-circuited. The identified parameters are then used for the description of the dynamic case. The results suggest that the nonlinear dynamic effects are mainly based on nonlinear hysteretic stress–strain behavior.

Zusammenfassung

Piezokeramische Sensoren und Aktoren wurden in den letzten Jahrzehnten in zahlreichen Bereichen der Technik angewendet. Im Bereich von kleinen Dehnungen, die aus schwachen elektrischen oder mechanischen Belastungen resultieren, wird das Verhalten von Piezokeramiken in der Regel durch lineare konstitutive Gleichungen beschrieben. Nichtlineare hysteretische Modelle werden verwendet, um den Polarisationsprozess oder das Verhalten von Piezokeramiken unter starken elektrischen Feldern oder mechanischen Spannungen oberhalb der koerzitiven Feldstärken zu beschreiben, wobei Orientierungsvorgänge der Domänen vorkommen.

Auf der anderen Seite kann nichtlineares Verhalten, vergleichbar mit einem degressiven Duffing-Oszillator, einschließlich Sprungphänomenen oder mehrfachen stabilen Lösungen bei gleicher Erregerssspannungsamplitude und -frequenz beobachtet werden, wenn polarisierte Piezokeramiken durch schwache elektrische Felder weit unterhalb der koerzitiven Feldstärke angeregt werden. Im Folgenden werden diese als dynamische Experimente bezeichnet. Die vorliegende Arbeit konzentriert sich auf die Beschreibung der nichtlinearen Effekte in Bereichen von mäßigen Dehnungen, wie sie typischerweise bei der Resonanzanregung solcher schwach gedämpfter Systeme bei dynamischen Untersuchungen vorkommen. Diese nichtlinearen Effekte können klassischerweise durch die Einführung von nichtkonservativen Termen und Termen höherer Ordnung in die elektrische Enthalpiedichte beziehungsweise in die konstitutiven Gleichungen beschrieben werden. Mit den Amplitudenfrequenzgängen aus dynamischen Versuchen nahe der Resonanz können die Parameter von Piezokeramiken bestimmt werden. Es ist aber schwierig, bei den nichtlinearen Kenngrößen zu entscheiden, welche Art der konservativen (mechanischen, piezoelektrischen oder dielektrischen) Nichtlinearität oder Dämpfung vorliegt.

Um diese Probleme zu bewältigen werden quasistatische Versuche mit angelegten mäßigen elektrischen Feldern sowie Zug- und Druckversuche bei mäßigen Spannungen, die in Dehnungen in der gleichen Größenordnung wie im dynamischen Fall resultieren, durchgeführt. Transversal polarisierte Piezokeramiken unter mäßigem quasistatischem elektrischem Feld in der Polarisationsrichtung weisen nichtlineare Hysteresen zwischen der Längsdehnung oder der elektrischen Verschiebungsdichte und den angelegten Feldern auf. Spannungs-Dehnungs-Hystereseverhalten wird auch bei Zug- und Druckversuchen beobachtet. Dieses quasistatische Verhalten kann dann durch vier der gängigsten Hysteresemodelle beschrieben werden, nämlich durch das klassische Preisach-Modell, das Prandtl-Ishlinskii-Modell, das Masing-Modell und das Bouc-Wen-Modell, die miteinander verknüpft sind.

Aufgrund des Vorteils, dass die Masing- und Bouc-Wen-Modelle durch die Differentialevolutionsgleichungen mit den inneren Variablen beschrieben werden, können diese Hysteresemodelle in die Modelle von Längsschwingungen der Piezokeramiken leicht eingebunden werden. Schließlich können die mechanischen Nichtlinearitäten direkt aus den Ergebnissen der Zug- und Druckversuche, unter der Bedingung, dass die Elektroden der Piezokeramik kurzgeschlossen sind, bestimmt werden. Die identifizierten Parameter werden dann für die Beschreibung des dynamischen Falls verwendet. Die Ergebnisse weisen darauf hin, dass die nichtlinearen dynamischen Effekte wesentlich auf dem nichtlinearen hysteretischen Spannungs-Dehnungs-Verhalten basieren.

Acknowledgments

The most sincere thanks to my supervisor, Prof. Dr.-Ing. Utz von Wagner, without his thoughtful guidance and enthusiastic support this study would not have been completed on time. I am greatly grateful to Prof. Dr. Peter Hagedorn for his fine reviews during the course of this work and for this dissertation. I am also deeply thankful to Prof. Dr. rer. nat. Wolfgang H. Müller for precious discussions about experiments and allowing me to use the microforce testing system in his laboratory. Once again, I would like to thank the three Professors as members of my thesis committee for providing invaluable comments and generous assessments.

This research was mainly funded by the Ministry of Education and Training of Vietnam (MOET) through Decision No. 3623/QĐ-BGDĐT on 09/7/2007. I am also grateful to the Deutscher Akademischer Austauschdienst (DAAD) for further fellowships, especially for the German course. I wish to express my gratitude to Prof. Dr.-Ing. Wilfried Kalkner and his coworkers, Dipl.-Ing. Christian Balkon and Mr. Torsten Haschke, for highly valuable technical support performing experiments with high voltages. The achievement of this work would not have been possible without the kind help of the following technical colleagues, Karl Theet, Ronald Koll and Wolfgang Griesche. Special thanks to Arion Juritza for spending a lot of time and effort on the experiments that contributed to the present work. I would like to thank Dr.-Ing. Daniel Hochlenert for fruitful discussions. I am also grateful and appreciative for all the help of other colleagues, who make my time at the Chair of Mechatronics and Machine Dynamics not only pleasant but also a learning experience. To Gisela Glass, thanks a lot for her friendly help and the warm and exciting atmosphere she creates. My gratitude goes to Nikolas Jüngel for his gracious help, especially when I first set foot in Berlin, and for the nice working atmotsphere in our office. I acknowledge to Guido Harneit, Stefan Schlagner, Holger Gödecker, Wolfram Martens, Kerstin Kracht, Sylwia Hornig, Alexander Lacher, and Yinuo Shi, who have offered friendly help in various ways.

In Vietnam, I am indebted to Prof. Dr.-Ing. Nguyễn Văn Khang (Hanoi University of Science and Technology) for leading me to the field of technical vibrations and recommending me to Prof. von Wagner. I would like to show my gratitude to the leadership of Hanoi University of Mining and Geology for their support. Special thanks to Dr. Trần Đình Sơn for his enthusiastic help and encouragement. To many friends and other people, who are not mentioned here but contribute to the success of this work, I also owe my deepest gratitude.

Last but certainly not least, I would like to devote my heartfelt thanks to my beloved paternal grandfather, my parents, my brother and his family for all the best they have done for me.

Contents

Abstract   i

Zusammenfassung   ii

Acknowledgments   iv

1 Introduction   1

2 Fundamentals of piezoelectricity   5
   2.1 Piezoelectric effects 5
   2.2 Linear theory of piezoelectricity 10
      2.2.1 Field quantities 10
      2.2.2 Mechanical equations 10
      2.2.3 Electrical equations 11
      2.2.4 Hamilton's principle for a piezoelectric solid 11
      2.2.5 Linear constitutive equations 12
      2.2.6 The system of linear equations of piezoelectricity 13

3 Dynamic experiments   15
   3.1 Experimental setup 15
   3.2 Nonlinear dynamic behavior 17

4 Linear dynamic modeling   21
   4.1 Linear constitutive equations 21
   4.2 Linear equations of motion 22
   4.3 Eigenfrequencies and eigenfunctions 23
   4.4 Ritz discretization 24
   4.5 Linear damping 26

5 Nonlinear dynamic modeling   27
   5.1 Nonlinear constitutive equations 28
   5.2 Ritz discretization 29
   5.3 Solution by perturbation analysis 30
   5.4 Determination of parameters 33
      5.4.1 Parameter identification from linear behavior 38
      5.4.2 Parameter identification from nonlinear behavior 40
   5.5 Consideration for quadratic nonlinearities 46

6 Quasi-static experiments 55
6.1 Experiments with moderate electric field 55
6.2 Experiments with moderate mechanical stress 59

7 Nonlinear quasi-static modeling 65
7.1 Hysteresis models 66
    7.1.1 Classical Preisach model 66
    7.1.2 Prandtl-Ishlinskii model 74
    7.1.3 Masing model 80
    7.1.4 Bouc-Wen model 81
7.2 Modeling of piezoceramics with hysteresis 82
    7.2.1 Piezoceramics under moderate electric field 83
    7.2.2 Piezoceramics under moderate mechanical stress 93
    7.2.3 Comparision of hysteresis models 102

8 Combination of nonlinear modelings 103
8.1 Dynamic modeling with Masing hysteresis 103
8.2 Dynamic modeling with Bouc-Wen hysteresis 106
8.3 Dynamic modeling with variable mechanical parameters 108

9 Conclusion and outlook 119

Bibliography 121

Chapter 1
Introduction

In recent decades, piezoceramics have found broad fields of applications. On the one hand, these smart materials, which possess piezoelectric properties, have been used for sensor applications, e.g. in accelerometers, microphones, load cells or rather for damping and power harvesting purposes by exploiting the direct piezoelectric effect [4]. On the other hand, the inverse piezoelectric effect can be used in actuators, for example polarized piezoceramics are bonded to the stator of travelling wave ultrasonic motors (USM), which are used e.g. to adjust the lens in the autofocus camera. The piezoceramics produce bending waves in the stator and the vibrations of the stator are then transmitted to the rotor by friction at contact points [42]. The stator of USM can also be fully made of piezoceramic material [104]. In automobile technology, piezoceramic stack actuators are used to control a needle valve that opens and closes a nozzle in order to spray fuel in the cylinder of an internal combustion engine [83]. Actuation systems of aircraft or helicopters based on piezoelectricity may gradually substitute conventional hydraulic systems [22]. In a recent project of NASA (National Aeronautics and Space Administration, USA) a full-scale helicopter with rotor blades containing piezoelectric stack actuators controlling the flap were tested. The results show that the system significantly reduced vibrations, saved energy and controlled rotor movement more precisely [105]. Piezoceramics are even integrated into sport equipments, such as snowboards [114, 115], bicycle structures [73] or golf clubs [43], where the materials undertake both energy harvesting and the function of dampers. Further applications of piezoelectric actuators, e.g. in information technology, robotics, bio and medical applications or ecological and energy applications can be found in [112].

The broad use of piezoceramic actuators is based on the number of advantages especially compared to pneumatic and hydraulic actuators in micro electro mechanical systems [24]. According to the manufacturer PI Ceramic, piezoceramic actuators can produce changes of position with subnanometer resolution and generate (or bear) forces of several tons. The response time of piezoceramic actuators is in range of microsecond and accelerition of 10,000 g can be reached. Electrical energy is only absorbed during dynamic operation of piezoceramic actuators and there is even no power consumption holding strong loads. Piezoceramic actuators are compatible with vacuum and cleanroom applications due to no use of lubricants or no abrasion. However, piezoelectric actuators also have several limitations, such as for the stack actuator the applied electric field is only allowed to be in the range from -0.3 to 1.5 kV/mm, tension forces and bending or torsion moments have to be avoided. Dynamic excitation results in dielectric and mechanical losses, which are nonlinearly related to the excitation frequency and amplitude or the humidity. The temperature in the actuators must typically be in the range from -40°C to 80°C [56]. Although only giving rise to small displacement, piezoceramic actuators exhibit nonlinear response relation with hysteresis and creep effects.

It is known that piezoceramics subjected to strong electric fields and/or mechanical stresses reveal a nonlinear behavior, such as hysteresis relations between the applied electric fields and spontaneous polarizations or strains, as well as creep phenomena. This is the case in the manufacturing processes to polarize the ceramic materials or the case of micro positioning where large displacements are required. Dealing with these nonlinearities, there have been numerous publications [1, 7, 8, 25, 51, 62–64, 77, 84, 93, 102, 103, 124, 130, 133], partially with specific applications and using different hysteresis models. These problems are often combined with other ones such as degradation of piezoelectric properties with respect to time [26] or combination with higher-order terms in the constitutive equations or energy functions [101, 102, 109]. Since these nonlinearities usually have a detrimental influence on the performance of piezoceramics in position and control applications, the description and solution of corresponding problems are taken interest in by a lot of works, for example [10, 21, 35, 57, 67, 74, 85, 94, 129].

These above-mentioned nonlinear effects can be accounted for by so-called polarization switching processes initiated when the electrical and/or mechanical loads reach sufficient coercive magnitude. On the other hand, in the range of very small strains resulting from weak electric fields or mechanical stresses, the behavior of piezoceramics can be described by linear constitutive equations in company with Newton's second law of motion, the linear strain–displacement relations as well as Maxwell's equations from the electric part. This linear theory can be found in a large number of textbooks, e.g. [9, 19, 47, 52, 79, 80, 86, 99, 111]. However, in some applications, for instance in piezo–beam systems [120] or ultrasonic travelling wave motors [104], piezoceramics are often excited near resonance by weak electric fields, under which switching processes may occur but are expected not to play a dominant role since the necessary magnitude of electrical or mechanical loads is not reached. Even for amplitude of electric fields in the range of 1–10 V/mm piezoceramics distinctly exhibit typical nonlinear vibration effects such as jump phenomena, multiple stable responses at the same excitation and the presence of superharmonics in spectra with monofrequent excitation. On the one hand, nonlinear effects can be taken into account by using nonlinear strain–displacement relations of von Kármán type as described e.g. in [58–60, 97, 98]. In the present work, displacements and strains are sufficiently small so that the linear mechanical relations can approximately be applied. On the other hand, the nonlinearities were dealt with by lots of authors starting with [11, 12], where higher-order elastic and dielectric terms were introduced into the uncoupling constitutive equations. Nonlinear energy functions, e.g. electric enthalpy and internal energy, were generally formulated in [79, 80]. This method was applied in several works, also with nonlinear boundary conditions [2, 5, 61, 89–92, 117–122, 128]. Energy harvesting applications based on piezo–beam systems regarding to such nonlinearities were also investigated [72, 108, 113].

In addition, investigations of nonlinear behavior of piezoceramics under strong mechanical stresses were done by performing tension and compression tests on both un-

Chapter 1. Introduction

polarized as well as polarized piezoceramics [31–34, 131]. The results obviously show asymmetric stress–strain behavior. Difference in plastic strains can be observed for polarized and unpolarized materials or for the same stress in tension and compression. A depolarization will occur in polarized piezoceramics subjected to tension stresses perpendicular to the polarization direction.

The aim of this work is to describe in more detail nonlinear dynamic effects caused by loads resulting in ranges of moderate strains occurring during the resonance operation of piezoceramic actuators excited by weak external electric fields far away from coercive ones, e.g. up to 3% of coercive electric field, in the case of the inverse 31-effect. According to the method of von Wagner [117] introducing higher-order conservative and nonconservative terms into the piezoelectric constitutive equations gives rise to good qualitative and quantitative accordance with experimental results. However, several questions remain open, for example for the type of nonlinearities, i.e. mechanical or coupling nonlinearities and quadratic or cubic ones, which plays a dominant role in the explanation of the nonlinear effects and for the model of the damping. Therefore, quasi-static experiments are performed with moderate electrical or mechanical loads resulting in moderate strain of the same order as in the dynamic experiments. The observed hysteretic behavior can be simulated by using different hysteresis models. Compatible models are then integrated into the linear dynamic modeling in order to interpret the dynamic nonlinearities. In another way, mechanical nonlinear parameters are determined directly from experimental hysteretic results and introduced into the nonlinear dynamic modeling. The corresponding results should be compared with those of experiments.

The present work is organized in nine chapters. In chapter 1 the problem statement, literature review and objective of the work are briefly given. Results in chapters 2–5 basically follow the work [117]. The fundamentals of piezoelectricity are described in chapter 2 introducing the direct and inverse piezoelectric effects in combination with the polarization process. Here the linear piezoelectric theory is also found including the thermodynamic energy function for the application of Hamilton's principle for piezoelectric continua, the linear piezoelectric constitutive equations, the mechanical equilibrium equations, the strain–displacement relations and quasi-electrostatic Maxwell's equations. Chapter 3 presents dynamic experiments with transversally polarized piezoceramics excited near the first resonance to longitudinal vibrations with respect to the inverse 31-effect. Various nonlinear effects can be observed at weak electric fields but resulting in moderate strains. In chapter 4 a linear modeling for longitudinal vibrations of the piezoceramics taking account of dissipative effects in the constitutive equations is considered. Chapter 5 contains the nonlinear modeling for the longitudinal vibrations describing the nonlinear effects observed in the dynamic experiments, where the quadratic and cubic nonlinearities are taken into account. The introduced parameters are then determined by fitting the experimental displacement amplitude responses. An additional consideration for quadratic nonlinearities leads to the ambigious problems of decision on the dominant type of nonlinearities or on modelling of the damping. In chapter 6 quasi-static experiments with transversally polarized piezoceramics excited by moderate electric fields using the inverse 31-effect or by moderate longitudinal mechanical stresses are described. The piezoceramics exhibit nonlinear hysteretic behavior of strain as well as electric displacement density to the excitations respectively. Chapter 7 is devoted to the modeling for these nonlinear hysteretic relations, where four of the most common hysteresis models are used. The corresponding parameters of the models are identified by using the experimental hysteresis curves. The interesting results of the combination of the dynamic and hysteretic modelings are presented in chapter 8. As above-mentioned, one possibility is integrating the hysteresis models into the linear conservative dynamic modeling and the other one is introducing the parameters identified from the stress–strain hysteretic relations into the nonlinear dynamic modeling. This work finishes with a conclusion and an outlook in chapter 9.

Chapter 2

Fundamentals of piezoelectricity

2.1 Piezoelectric effects

The (direct) piezoelectric effect was experimentally discovered in 1880 by the French physicists Jacques Curie and Pierre Curie [19, 52]. The brothers demonstrated that compressing certain crystals, for example those of tourmaline, quartz or Seignette's salt, generates internally electrical charges which results in an electric field inside the crystal [23]. In the following year, the inverse effect predicted from thermodynamic consideration by Lippmann [76], that the crystals is deformed when subjected to an external electric field, was also verified by the brothers Curie. Due to the fact that the above natural monocrystalline materials exhibit very weak piezoelectric effects, polycrystalline materials with enhanced properties have been developed. The most widely used materials in modern technical applications are barium titanate ($BaTiO_3$) and lead zirconate titanate (PZT) which is mainly synthesized from lead zirconate ($PbZrO_3$) and lead titanate ($PbTiO_3$). The samples used in this work are made of PZT, manufactured by PI Ceramic, Germany.

The genesis of both piezoelectric effects in barium titanate or PZT can be explained by considering an unit cell for example of barium titanate sketched in figure 2.1 [63]. Above the Curie temperature T_C ($BaTiO_3$: 120 – 130°C, PZT: 250 – 370°C) the unit cell is of cubic shape and the centers of positive and negative electric charges coincide. In this state, there is no permanent dipole within the unit cell, i.e. it possesses no spontaneous polarization. Below the Curie temperature T_C the unit cell becomes tetragonal, the centers of positive and negative charges no longer coincide. Hence, the unit cell now has a spontaneous polarization and consequently piezoelectric properties. Indeed, the unit cell subjected to a mechanical stress is deformed leading to a relative displacement of the centers of the positive and negative charges. This changes the polarization or rather the electric state. On the other hand, when an electric field is applied to the unit cell, if the direction of the field is the same as that of the spontaneous polarization, then the centers of the charges move apart from each other and vice versa for an opposite electric field. This movement brings about a corresponding elongation or contraction of the unit cell.

In the manufacturing operations, the piezoceramic materials (barium titanate or PZT) suffer a sintering at temperatures above the Curie temperature followed by a subsequent cooling to temperatures below the Curie temperature [6, 63]. During the

Chapter 2. Fundamentals of piezoelectricity

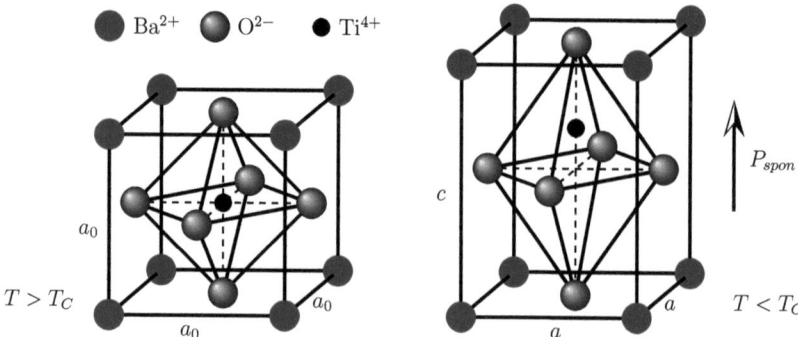

Figure 2.1: Unit cell of barium titanate (BaTiO$_3$). Left: cubic cell at temperature above Curie temperature ($T > T_C$). Right: tetragonal cell below Curie temperature ($T < T_C$) with a shift of the centers of the positive and negative electric charges resulting in spontaneous polarization P_{spon}. At room temperature $c \approx 1.01a$ [55, 63].

cooling process, the central cation of each unit cell can move along one of three axes of the original cube, giving six different possibilities of the spontaneous polarization. Therefore, the spontaneous polarization of the unit cells is not unique over a grain with specific orientation of crystal lattice but distributed randomly with homogeneous direction only in substructures of the grain, called Weiss domains. Due to this random distribution, the spontaneous polarization of the domains cancel macroscopically each other conducting the thermally depoled state of the material, in which the macroscopic continuum exhibits no piezoelectricity.

For the use in technical applications as sensors or actuators the thermally depoled piezoceramic materials have to be polarized, so that they possess piezoelectric properties. The polarization process is performed by applying a strong electric field with a magnitude above the coercive field E_c, at which the spontaneous polarizations of the domains begin switching. The manufacturer PI Ceramic gives the value of the coercive electric field as 1.2 kV/mm and 1.5 kV/mm in case of the PZT materials PIC 255 and PIC 181 respectively. After removal of the poling electric field, the ceramic has a remanent polarization in the direction of the applied electric field named z- or 3-direction. Because the spontaneous polarization of some domains may not switch to coincide perfectly with the direction of the field, the actual value of the remanent polarization can only reach to the so-called saturation polarization $P_{sat} \leq P_{spon}$. The orientation of the macroscopic polarization is accompanied by a remanent strain of the whole material as a result of the spontaneous elongations of the unit cells. The remanent strain gets a maximum value called saturation strain S_{sat} which can not exceed the strain corresponding to the spontaneous polarization.

Applying an external electric field to the free polarized piezoceramic body in the 3-direction results in an elongation or contraction in the same direction (depending on the direction of the external field), called the inverse 33-effect which relates to the piezoelectric coefficient d_{33}, in company with a corresponding lateral strain of opposite

Chapter 2. Fundamentals of piezoelectricity

sign, named the inverse 31-effect, for which the piezoelectric coefficient d_{31} is relevant. Figure 2.2 illustrates these effects for a piezoceramic rod subjected to a constant electric voltage. On the other hand, an applied electric field in the transverse (1-)direction perpendicular to the polarization direction gives rise to a shear strain in the 13-plane, called the inverse 15-effect with respect to the piezoelectric coefficient d_{15}, where the index "13" is abbreviated by "5" as later shown in table 2.1. All three inverse effects are used for piezoceramic actuators including e.g. stack, bending or shear actuators. The direct piezoelectric effects used in sensors are obtained when mechanical stresses are applied to the piezoceramics resulting in measurable electrode potential.

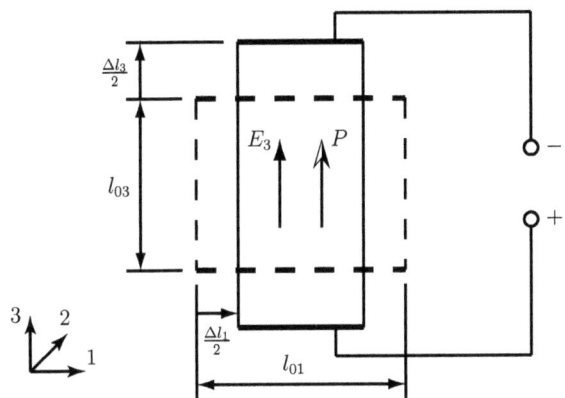

Figure 2.2: Deformation of piezoceramics under an electric field in the polarization direction: the inverse piezoelectric 33- and 31-effects.

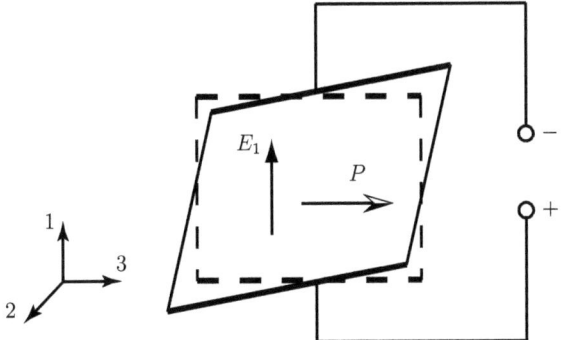

Figure 2.3: Deformation of piezoceramics under an electric field perpendicular to the polarization direction: the inverse piezoelectric 15 effect.

For a thorough understanding of the piezoelectric effects, the so-called dielectric hysteresis and butterfly hysteresis, which are directly related to the polarization pro-

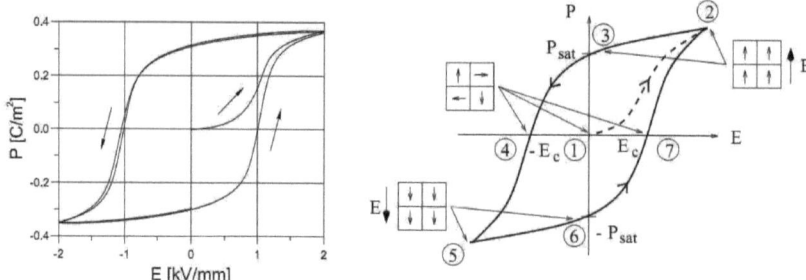

Figure 2.4: Dielectric hysteresis [63]. Left: polarization P vs. applied electric field E for PIC 151. Right: schematic sketch of dielectric hysteresis.

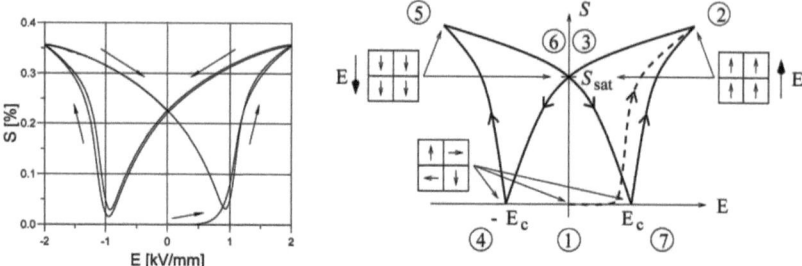

Figure 2.5: Butterfly hysteresis [63]. Left: strain S vs. applied electric field E for PIC 151. Right: schematic sketch of butterfly hysteresis.

cess, respectively shown in figure 2.4 and 2.5 will be described. This was presented in the excellent review article of Kamlah [63].

Experiments with free thermally depoled piezoceramic specimens of the material PIC 151 were done by applying a strong uniaxial electric field and recording polarization and strain in the direction of the field. Starting from the state ① with the random orientation of the polarization of the domains and raising the strength of the electric field to the coercive value E_c initiates the switching processes of the domains. In the state ② corresponding to $E = 2$ kV/mm almost uniform orientation of the domains is reached. Reducing the electric field to zero attains the polarized state ③, where the piezoceramic is then ready for utilization.

The electric field is now applied in opposite direction and reaching the coercive field $-E_c$ in the state ④ will theoretically depolarize the piezoceramics. Here the macroscopic polarization also vanishes but the distribution of the domains may be different from that in the state ①. Next, the piezoceramics could be repolarized by raising the electric field strength up to 2 kV/mm again then withdrawing the field as from the state ④ to ⑥. A new reversing of the applied electric field results in the responses of the material in the similar way that follows the states ⑥ - ⑦ - ②.

Chapter 2. Fundamentals of piezoelectricity 9

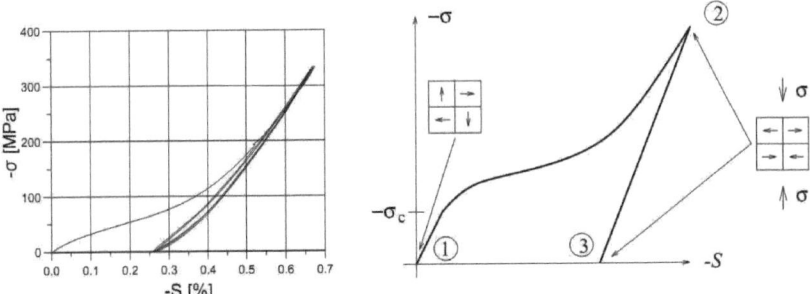

Figure 2.6: Ferroelastic hysteresis [63]. Left: applied compressive stress $-\sigma$ vs. compressive strain $-S$ for PIC 151. Right: schematic sketch of ferroelastic hysteresis.

In the present work, the nonlinear behavior of the polarized piezoceramics in the vicinity of a so-called working point corresponding to the states ③ or ⑥ will be considered. This is normally the main operating region of piezoceramics in applications.

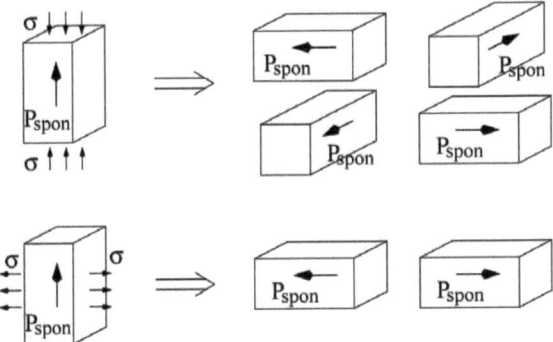

Figure 2.7: Polarization switching possibilities under strong mechanical stresses [63].

Also according to Kamlah [63] switching processes can be initiated by strong mechanical stresses as well. In experiments, thermally depoled piezoceramics of the material PIC 151 were subjected to strong uniaxial compressive stresses in three loading–unloading cycles. The so-called ferroelastic response of the material and the corresponding schematic sketch are shown in figure 2.6. For small stress near to the starting point ① the material exhibits a linear elastic behavior. Reaching the coercive stress σ_c starts polarization switching processes of the domains. There are four different possibilities for the compressive stress to switch an unit cell by 90° as sketched in figure 2.7 (top), where no direction is preferred. The piezoceramics now show a nonlinear stress–strain relation. The fully switched domain structure is derived in the state ②. Due to the random distribution of the spontaneous polarizations of the domains in the plane perpendicular to the compressive load, the domain state is now isotropic in that plane. Therefore, the piezoceramics possess no macroscopic remanent polarization. This domain structure is essentially conserved when the load is reduced to zero in the state ③ as well as in next loading–unloading cycles. In this work, tension and compression test are carried out, but the magnitude of the mechanical loads is only about 10% of that mentioned above and the hysteretic stress–strain relation is macroscopically of interest.

2.2 Linear theory of piezoelectricity

2.2.1 Field quantities

In the theory of piezoelectricity, both the mechanical and electrical quantities are considered, including

$\mathbf{T} = (T_{ij}), \quad i,j = 1,2,3 \quad \text{or} \quad i,j = x,y,z :$ Stress tensor,
$\mathbf{S} = (S_{ij}), \quad i,j = 1,2,3 \quad \text{or} \quad i,j = x,y,z :$ Strain tensor,
$\vec{u} = (u_x, u_y, u_z) \quad \text{or} \quad \vec{u} = (u,v,w) :$ Mechanical displacement vector,
$\vec{D} = (D_i), \quad i = 1,2,3 \quad \text{or} \quad i = x,y,z :$ Electric displacement field vector,
$\vec{E} = (E_i), \quad i = 1,2,3 \quad \text{or} \quad i = x,y,z :$ Electric field vector,
$\varphi \quad :$ Scalar electric potential.

The field quantities are coupled in the piezoelectric constitutive equations. Newton's second law and the strain–displacement equations are used for the mechanical behavior and Maxwell's equations are used for the electrical one.

2.2.2 Mechanical equations

Deriving from Newton's second law, the equations of motion for a continuum can be expressed in Cartesian coordinates as

$$\frac{\partial T_{xx}}{\partial x} + \frac{\partial T_{xy}}{\partial y} + \frac{\partial T_{xz}}{\partial z} + F_x = \rho \frac{\mathrm{d}}{\mathrm{d}t}(\dot{u}_x), \tag{2.1}$$

$$\frac{\partial T_{xy}}{\partial x} + \frac{\partial T_{yy}}{\partial y} + \frac{\partial T_{yz}}{\partial z} + F_y = \rho \frac{\mathrm{d}}{\mathrm{d}t}(\dot{u}_y), \tag{2.2}$$

$$\frac{\partial T_{xz}}{\partial x} + \frac{\partial T_{yz}}{\partial y} + \frac{\partial T_{zz}}{\partial z} + F_z = \rho \frac{\mathrm{d}}{\mathrm{d}t}(\dot{u}_z), \tag{2.3}$$

where F_i and ρ are respectively the body forces per unit of volume acting on every point within the body and the constant mass density of the material. These equations can be more shortly rewritten using Einstein summation convention as

$$T_{ij,j} + F_i = \rho \frac{\mathrm{d}}{\mathrm{d}t}(\dot{u}_i), \tag{2.4}$$

where $()_{\bullet,j}$ is shorthand for the partial derivative with respect to the coordinates $j = 1,2,3$ or $j = x,y,z$. The strain–displacement equations represent the relation

Chapter 2. Fundamentals of piezoelectricity

between the displacement and the strain

$$S_{xx} = \frac{\partial u_x}{\partial x}, \quad S_{yy} = \frac{\partial u_y}{\partial y}, \quad S_{zz} = \frac{\partial u_z}{\partial z}, \quad (2.5)$$

$$S_{xy} = \frac{1}{2}\left(\frac{\partial u_x}{\partial y} + \frac{\partial u_y}{\partial x}\right), \quad S_{xz} = \frac{1}{2}\left(\frac{\partial u_x}{\partial z} + \frac{\partial u_z}{\partial x}\right), \quad S_{yz} = \frac{1}{2}\left(\frac{\partial u_y}{\partial z} + \frac{\partial u_z}{\partial y}\right), \quad (2.6)$$

which can be abbreviated as

$$S_{ij} = \frac{1}{2}(u_{i,j} + u_{j,i}). \quad (2.7)$$

2.2.3 Electrical equations

Because the phase speeds of electromagnetic waves are five orders of magnitude higher than the speeds of elastic waves [82, 111, 117], the quasi-electrostatic approximation can be used. This simplifying assumption reduces Maxwell's equations to

$$D_{i,i} = \rho_f, \quad (2.8)$$
$$E_i = -\varphi_{,i}, \quad (2.9)$$

where ρ_f denotes the free charge density. Piezoceramics will be considered as polarizable dielectrics only, so the free charge density $\rho_f = 0$. Therefore,

$$D_{i,i} = 0. \quad (2.10)$$

The equations describing the relation between mechanical and electrical fields can be formulated using the principle of conservation of energy as well as Hamilton's principle for a piezoelectric medium [127].

2.2.4 Hamilton's principle for a piezoelectric solid

In the absence of volume forces, Hamilton's principle for a piezoelectric continuum in a volume V bounded by a surface A is expressed as [127]

$$\delta \int_{t_0}^{t_1} L(u_i, \varphi)\, dt + \int_{t_0}^{t_1} \delta W(u_i, \varphi)\, dt = 0. \quad (2.11)$$

In the equation (2.11), the Lagrangian L is defined by

$$L = \int_V \left(\frac{1}{2}\rho \dot{u}_i^2 - H\right) dV = \int_V (T - H)\, dV, \quad (2.12)$$

where ρ denotes the constant mass density, T and H correspond to the kinetic energy density and the electric enthalpy density respectively. In addition, the virtual work done by the nonconservative surface forces in a virtual displacement (δu_i) of the surface and the electrical analog of the virtual work done by the surface charges in a variation of electric potential ($\delta \varphi$) are given as

$$\delta W = \int_A (\bar{f}_i \delta u_i - \bar{\sigma}\, \delta \varphi)\, dA, \quad (2.13)$$

with $\bar{f}_i$ and $\bar{\sigma}$ are respectively the prescribed surface forces and surface charge density on some part of the surface A.

2.2.5 Linear constitutive equations

Starting from the principle of conservation of energy and considering the independent variables S_{ij} and E_i, the electric enthalpy density in a linear theory can be constructed in a homogeneous quadratic form as [111]

$$H = \frac{1}{2} c^E_{ijkl} S_{ij} S_{kl} - e_{ijk} E_i S_{jk} - \frac{1}{2} \varepsilon^S_{ij} E_i E_j, \qquad (2.14)$$

so that the following conditions are satisfied

$$T_{ij} = \frac{\partial H}{\partial S_{ij}}, \qquad D_i = -\frac{\partial H}{\partial E_i}. \qquad (2.15)$$

Then the linear piezoelectric constitutive equations can be derived as

$$\begin{aligned} T_{ij} &= c^E_{ijkl} S_{kl} - e_{kij} E_k, & (2.16) \\ D_i &= e_{ikl} S_{kl} + \varepsilon^S_{ik} E_k, & (2.17) \end{aligned}$$

where c^E_{ijkl}, e_{ijk} and ε^S_{ij} correspond to elastic, piezoelectric and dielectric constants. The superscripts E and S indicate the constant electric field and strain conditions for the parameter determination, respectively.

In case the stress T_{ij} and the electric field E_i are considered as independent variables, the constitutive equations can be obtained as [52, 61]

$$\begin{aligned} S_{ij} &= s^E_{ijkl} T_{kl} + d_{kij} E_k, & (2.18) \\ D_i &= d_{ikl} T_{kl} + \varepsilon^T_{ik} E_k, & (2.19) \end{aligned}$$

where s^E_{ijkl}, d_{ijk} and ε^T_{ij} correspond to elastic, piezoelectric and dielectric constants.

Due to the symmetry of the elastic and piezoelectric tensors, these equations can be expressed using a compressed matrix notation. The following identifications are first introduced [82]

$$\begin{cases} s^E_{ijkl} = s^E_{pq}, & i = j \wedge k = l \\ 2 s^E_{ijkl} = s^E_{pq}, & i = j \wedge k \neq l \\ 4 s^E_{ijkl} = s^E_{pq}, & i \neq j \wedge k \neq l \end{cases} \quad \text{and} \quad \begin{cases} d_{ikl} = d_{iq}, & k = l \\ 2 d_{ikl} = d_{iq}, & k \neq l \end{cases}, \qquad (2.20)$$

$$T_{ij} = T_p \quad \text{and} \quad \begin{cases} S_{ij} = S_p, & i = j \\ 2 S_{ij} = S_p, & i \neq j \end{cases}, \qquad (2.21)$$

where the new indices p and q are set according to table 2.1. The linear constitutive equations (2.18) and (2.19) now become

$$\begin{aligned} S_p &= s^E_{pq} T_q + d_{kp} E_k, & (2.22) \\ D_i &= d_{iq} T_q + \varepsilon^T_{ik} E_k. & (2.23) \end{aligned}$$

In the general case, these constitutive equations contain 21 independent elastic constants, 18 independent piezoelectric constants and 6 independent dielectric constants.

The piezoceramics used in this work show the behavior of an orthotropic material

Chapter 2. Fundamentals of piezoelectricity   13

ij or kl	p or q
11	1
22	2
33	3
23 or 32	4
31 or 13	5
12 or 21	6

Table 2.1: Tensor index abbreviations.

after polarization process, where in the 1-direction perpendicular to the polarization direction the piezoceramics exhibit an isotropic behavior. Therefore, the number of the material constants reduces to 5 elastic constants, 3 piezoelectric constants and 2 dielectric constants. The linear constitutive equations (2.22) and (2.23) can then be expanded as [127]

$$S_1 = s_{11}^E T_1 + s_{12}^E T_2 + s_{13}^E T_3 + d_{31} E_3, \tag{2.24}$$
$$S_2 = s_{12}^E T_1 + s_{11}^E T_2 + s_{13}^E T_3 + d_{31} E_3, \tag{2.25}$$
$$S_3 = s_{13}^E T_1 + s_{13}^E T_2 + s_{33}^E T_3 + d_{33} E_3, \tag{2.26}$$
$$S_4 = s_{44}^E T_4 + d_{15} E_2, \tag{2.27}$$
$$S_5 = s_{44}^E T_5 + d_{15} E_1 \tag{2.28}$$
$$S_6 = 2\left(s_{11}^E - s_{12}^E\right) T_6, \tag{2.29}$$
$$D_1 = d_{15} T_5 + \varepsilon_{11}^T E_1, \tag{2.30}$$
$$D_2 = d_{15} T_4 + \varepsilon_{11}^T E_2, \tag{2.31}$$
$$D_3 = d_{31} T_1 + d_{31} T_2 + d_{33} T_3 + \varepsilon_{33}^T E_3. \tag{2.32}$$

2.2.6 The system of linear equations of piezoelectricity

In summary, a determined system of equations of the linear theory of piezoelectricity is taken into account. This differential algebraic system consists of 22 equations in 22 variables [111]. The equations are three equilibrium equations (2.4) with respect to 6 components of the stress tensor and 3 components of the displacement vector with the neglected body forces, six equations for the strain–displacement relation (2.7) with 6 elements of the strain tensor as additional variables, four Maxwell's equations (2.9) and (2.10) with 7 additional variables, which are 3 elements of the electric displacement field vector, 3 elements of the electric field vector and the scalar electric potential, and nine linear constitutive equations (2.22) and (2.23).

Chapter 3
Dynamic experiments

In contrast to the linear behavior represented by the linear theory of piezoelectricity described in section 2.2, nonlinear effects can easily be observed in dynamic experiments for longitudinal vibrations of piezoceramics. Such experiments with transversally polarized piezoceramics will be presented in this chapter. In the dynamic experiments, piezoceramics are excited close to the first resonance by weak electric fields with respect to the inverse 31-effect. These experiments are performed according to the corresponding procedure described in [90, 117].

For experimental investigations transversally polarized piezoceramic rods of the materials PIC 255 and PIC 181 manufactured by PI Ceramic in Lederhose (Thüringen), Germany, are used. Due to possessing slight damping, the "hard" piezoceramic PIC 181 particularly exhibits more evident nonlinear effects.

3.1 Experimental setup

An overview of experimental setup used in this work for dynamic experiments under weak electric fields is shown in figure 3.1. A free piezoceramic rod, which is transversally polarized in the z-direction, is excited to longitudinal vibrations near the first resonance by harmonic excitation voltages using the inverse 31-effect, this means the applied electric field in the polarization direction z is transversal to the main vibration direction x. Experiments were performed with variety of piezoceramic samples made of the material PIC 255 and PIC 181. The amplitude and phase of the displacements at one end of the piezoceramic rods can be obtained by means of the experimental setup sketched in figure 3.2. The actual experimental setup is presented in figure 3.3.

In fact, the piezoceramics are excited near the first eigenfrequency by using an gain-phase analyzer (Hewlett-Packard 4192A LF). This gain-phase analyzer generates a sweep up and sweep down of excitation frequency and processes the response signal. The excitation signal is magnified by a power amplifier (Brüel & Kjær 2713) and then applied to the piezoceramics. This output of the amplifier is simultaneously fed to the gain-phase analyzer as a reference input. An oscilloscope (Gould DSO 1504) and an integrating digital multimeter (Prema 6030) are used to monitor the excitation voltages and responses. Data from the gain-phase analyzer are finally transfered to a measuring computer through an USB/GPIB-Interface cable (Aligent Technologies 82357A).

16   Chapter 3. Dynamic experiments

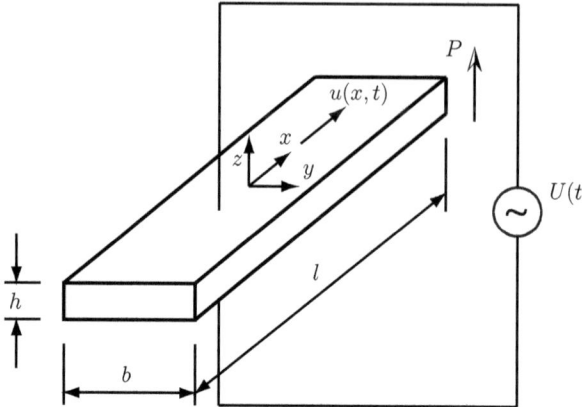

Figure 3.1: Polarized piezoceramic rod subjected to electric voltage.

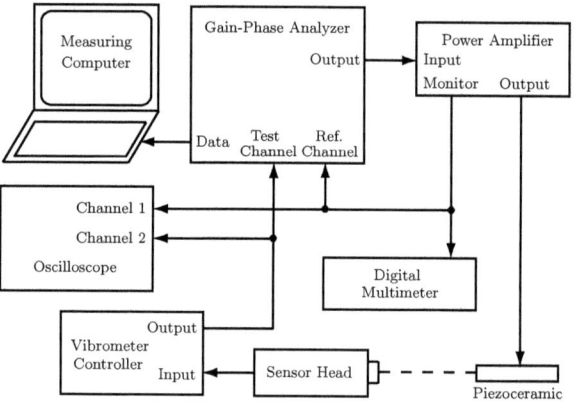

Figure 3.2: Schematic experimental setup of dynamic experiments.

The responses of the piezoceramics can be derived as velocities measured by two units of a laser vibrometer (Polytec). The laser beam from the sensor head (OFV-302) is directed at the surface of the piezoceramics. The reflected signals are processed by the modular controller (OFV-3000) and passed to the test channel of the gain-phase analyzer. The output of the vibrometer is recorded in the measuring computer and also monitored by the oscilloscope.

The gain-phase analyzer produces a measurement for gain and phase shift between the excitation signal and the response signal by comparing the reference channel with the test one. The excitation voltage can be regarded as an ideal harmonic. There exist superharmonics in the responses, but near the first resonance they are in much smaller order of magnitude in comparision with the part corresponding to the excitation fre-

Chapter 3. Dynamic experiments

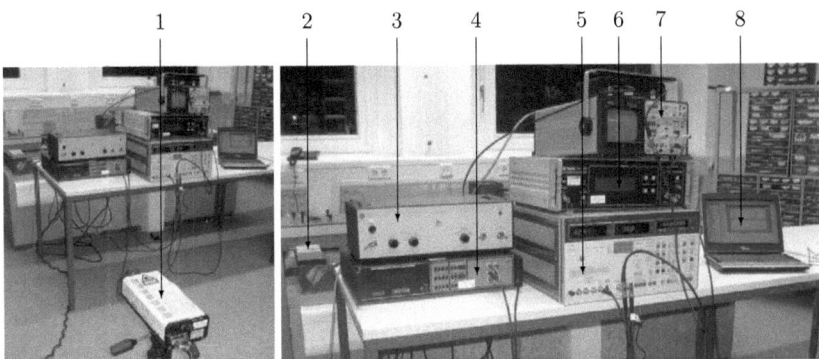

Figure 3.3: Experimental setup of dynamic experiments: sensor head (1), polarized piezoceramic rod located on a soft foam foundation (2), power amplifier (3), digital multimeter (4), gain-phase analyzer (5), vibrometer controller (6), oscilloscope (7) and measuring computer (8).

quency. Therefore the response signals can also be considered to be harmonic, so that the gain can be interpreted as amplitude behavior of the velocity to the excitation. Then from the amplitudes of the harmonic velocities, it is easy to calculate the corresponding displacement amplitudes at one end of the piezoceramics with the given excitation frequencies.

3.2 Nonlinear dynamic behavior

Dynamic experiments on samples of the material PIC 255 with the dimensions $70 \times 25 \times 3.3$ mm^3 ($l \times b \times h$) and those of the material PIC 181 with the dimensions $30 \times 3 \times 2$ mm^3 are performed. The first eigenfrequency is about 20 kHz for PIC 255 samples and 56 kHz for PIC 181 samples. Figures 3.4 and 3.5 show the displacement amplitude–frequency responses of PIC 255 and PIC 181 samples, respectively, near to the first resonance at various excitation voltages, where the displacement amplitude is normalized by the excitation amplitude. The following nonlinear phenomena can be observed from these experimental results:

- The first resonance frequency is reduced by higher excitation voltage.
- The normalized displacement amplitude is reduced by higher excitation voltage.
- There exist jumps of the displacement amplitude and different stable responses for the sweep up and sweep down of excitation frequency for the material PIC 181.

The experiments on the material PIC 181 also indicate that the approximately linear behavior can no longer be observed even with very small external applied electric field of about 1 V/mm. The aforementioned nonlinear effects are the same as those in [90,117], where they also occurred for the normalized electric current through the piezoceramics.

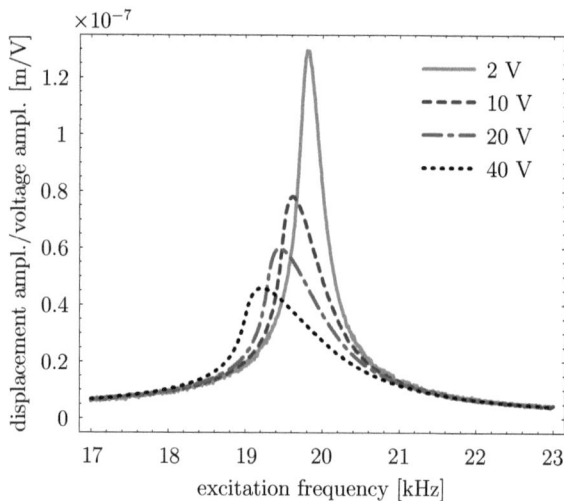

Figure 3.4: Normalized displacement amplitude response of PIC 255.

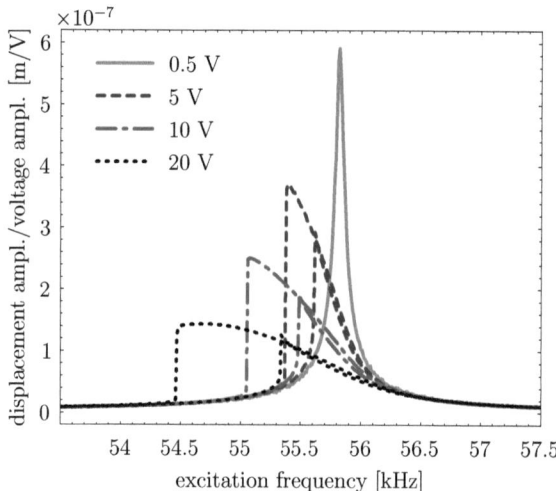

Figure 3.5: Normalized displacement amplitude response of PIC 181. The jumps on the right and on the left correspond to the sweep up and sweep down of excitation frequency respectively.

Chapter 3. Dynamic experiments

The form of the amplitude–frequency response curves shows a behavior of a conventional Duffing oscillator with softening cubic stiffness. However, this can also be accounted for by quadratic stiffness [87]. The reduction of the normalized resonance amplitude can be explained by nonlinear damping or nonlinear piezoelectric coupling effects [117]. In comparision with PIC 181, the "soft" piezoceramic material PIC 255 exhibits a low mechanical quality, or in other words, it is a material with higher damping.

Chapter 4
Linear dynamic modeling

In this chapter a linear modeling of the dynamic behavior of piezoceramics exhibited in the experiments described in chapter 3 is presented. This modeling was briefly given in [90, 117]. First the linear equation of motion is derived by using the linear theory of piezoelectricity described in section 2.2. From this equation, eigenfrequencies and eigenfunctions of free longitudinal vibrations of transversally polarized piezoceramic rods are obtained. Applying Ritz discretization to the equation of motion, an approximate solution can be found. Furthermore nonconservative effects are considered by introducing linear damping terms into the piezoelectric constitutive equations.

4.1 Linear constitutive equations

In order to derive a simple linear modeling for longitudinal vibrations of tranversally polarized piezoceramic rods, it can be assumed, that the transversal stresses, the shear stresses, the shear strains and the electric field in the directions perpendicular to the polarization direction are negligible, this means

$$T_{yy} = T_{zz} = T_{xy} = T_{yz} = T_{xz} \equiv 0, \quad S_{xy} = S_{yz} = S_{xz} \equiv 0, \quad E_x = E_y \equiv 0. \tag{4.1}$$

The linear piezoelectric constitutive equations (2.24)–(2.32) reduce to

$$S_{xx} = s_{11}^E T_{xx} + d_{31} E_z, \tag{4.2}$$
$$S_{yy} = s_{12}^E T_{xx} + d_{31} E_z, \tag{4.3}$$
$$S_{zz} = s_{13}^E T_{xx} + d_{33} E_z, \tag{4.4}$$
$$D_z = d_{31} T_{xx} + \varepsilon_{33}^T E_z. \tag{4.5}$$

Solving equation (4.2) for the stress T_{xx} and substituting this into (4.5) yields

$$T_{xx} = E^{(0)} S_{xx} - \gamma_0 E_z, \tag{4.6}$$
$$D_z = \gamma_0 S_{xx} + \nu_0 E_z, \tag{4.7}$$

where

$$E^{(0)} = \frac{1}{s_{11}^E}, \quad \gamma_0 = \frac{d_{31}}{s_{11}^E}, \quad \nu_0 = \varepsilon_{33}^T - \frac{d_{31}^2}{s_{11}^E}. \tag{4.8}$$

According to (2.15) and (4.1)

$$T_{xx} = \frac{\partial H}{\partial S_{xx}}, \quad T_{yy} = \frac{\partial H}{\partial S_{yy}} = 0, \quad T_{zz} = \frac{\partial H}{\partial S_{zz}} = 0, \quad D_z = -\frac{\partial H}{\partial E_z} \tag{4.9}$$

Hence, the respective electric enthalpy density H has to be a quadratic form as (2.14)

$$H = \frac{1}{2} E^{(0)} S_{xx}^2 - \gamma_0 S_{xx} E_z - \frac{1}{2} \nu_0 E_z^2. \qquad (4.10)$$

4.2 Linear equations of motion

Neglecting the volume force F_x and using the assumption (4.1), the equation of motion for the longitudinal vibrations of piezoceramic rods can be derived from Newton's second law (2.4) as

$$T'_{xx} = \rho \ddot{u}, \qquad (4.11)$$

where $()' = \partial()/\partial x$ and $u(x,t)$ is the displacement in longitudinal direction of the rod. Taking the constitutive equation (4.6) into account leads to

$$E^{(0)} S'_{xx} - \gamma_0 E'_z = \rho \ddot{u}. \qquad (4.12)$$

Substituting the strain from (2.5)

$$S_{xx} = u' \qquad (4.13)$$

and the electric field from (2.9), namely

$$E_z = -\frac{\partial \varphi}{\partial z} \qquad (4.14)$$

into the equation (4.12), the linear equation of motion is obtained as

$$\rho \ddot{u} = E^{(0)} u'' \qquad (4.15)$$

because from Maxwell's equation (2.10) and the constitutive equation (4.7)

$$\frac{\partial^2 \varphi}{\partial z^2} = 0. \qquad (4.16)$$

Since an alternating voltage $U(t) = U_0 \cos \Omega t$ is applied to the electrodes of the piezoceramic rod, the electric potentials at these electrodes can be written as

$$\varphi\left(z = \frac{h}{2}, t\right) = -\frac{U_0}{2} \cos \Omega t, \qquad (4.17)$$

$$\varphi\left(z = -\frac{h}{2}, t\right) = \frac{U_0}{2} \cos \Omega t. \qquad (4.18)$$

Then the electric potential is the solution of the boundary value problem (4.16)–(4.18)

$$\varphi = -\frac{U_0}{h} z \cos \Omega t \qquad (4.19)$$

and the electric field can be obtained from the equation (4.14) as

$$E_z = \frac{U_0}{h} \cos \Omega t, \qquad (4.20)$$

where U_0 is the excitation voltage amplitude, Ω is the excitation angular frequency and h is the thickness of the piezoceramic rod. Therefore, two boundary conditions can be

Chapter 4. Linear dynamic modeling

derived by appling the constitutive equation (4.6) to both free ends of the rod where the stress T_{xx} vanishes

$$E^{(0)}\, u'\left(-\frac{l}{2},t\right) = E^{(0)}\, u'\left(\frac{l}{2},t\right) = \gamma_0\, \frac{U_0}{2}\, \cos \Omega t. \tag{4.21}$$

The linear equation of motion (4.15), together with the dynamic boundary conditions (4.21), can also be derived by using Hamilton's principle as presented in [90].

4.3 Eigenfrequencies and eigenfunctions

The linear eigenfunctions and eigenfrequencies of the free longitudinal vibrations can be derived by solving the field equation (4.15) with the dynamic boundary conditions (4.21) in the case of vanishing excitation, i.e. $U_0 = 0$. This physically means, that the electrodes are short-circuited. The rewritten field equation is now considered as

$$\ddot{u} = c^2\, u'', \quad c = \sqrt{\frac{E^{(0)}}{\rho}}. \tag{4.22}$$

Unlike the case of the longitudinal vibrations of polarized piezoceramics with respect to the 33-effect described in [117], here the free vibrations of the transversally polarized piezoceramics are the same as those of a rod without piezoelectric properties. Therefore, assuming separation of variables, solutions will be sought of the product form

$$u(x,t) = U(x)\, p(t). \tag{4.23}$$

Both sinusoidal and cosine eigenfunctions can then be derived [41]. Indeed, the sinusoidal eigenfunctions are

$$U_k(x) = \sin\left[(2k-1)\,\pi\, \frac{x}{l}\right], \quad x \in \left[-\frac{l}{2}, \frac{l}{2}\right], \tag{4.24}$$

where l is the length of the rod and $k = 1, 2, 3, \ldots, \infty$. Each of the sinusoidal eigenfunctions belongs to an angular eigenfrequencies which can be determined as

$$\omega_k = \frac{(2k-1)\,\pi}{l}\sqrt{\frac{E^{(0)}}{\rho}}, \quad k = 1, 2, 3, \ldots, \infty. \tag{4.25}$$

On the other hand, the cosine eigenfunctions are

$$U_k^*(x) = \cos\left(2\,k\,\pi\, \frac{x}{l}\right), \quad x \in \left[-\frac{l}{2}, \frac{l}{2}\right] \quad \text{and} \quad k = 1, 2, 3, \ldots, \infty. \tag{4.26}$$

The corresponding angular eigenfrequencies can be calculated as

$$\omega_k^* = \frac{2\,k\,\pi}{l}\sqrt{\frac{E^{(0)}}{\rho}}, \quad k = 1, 2, 3, \ldots, \infty. \tag{4.27}$$

Here $\omega_0^* = 0$ means the displacement of the rod as a rigid boby, which is not considered.

However, the cosine eigenmodes of the longitudinal vibrations could not be excited by applying an electric field. Let us consider longitudinal vibrations of a piezoceramic rod excited by alternating voltage $U(t) = U_0 \cos \Omega t$. The form of solution

$$u(x,t) = U(x) \cos \Omega t \tag{4.28}$$

is used and this satisfies the field equation (4.22) if

$$U(x) = C_1 \cos \frac{\Omega}{c} x + C_2 \sin \frac{\Omega}{c} x, \tag{4.29}$$

where C_1 and C_2 are unknown constants. Substituting the solution of the field equation into the boundary conditions (4.21) yields

$$C_1 \sin \frac{\Omega}{c} \frac{l}{2} + C_2 \cos \frac{\Omega}{c} \frac{l}{2} = \gamma_0 \frac{U_0}{2 E^{(0)}} \frac{c}{\Omega}, \tag{4.30}$$

$$-C_1 \sin \frac{\Omega}{c} \frac{l}{2} + C_2 \cos \frac{\Omega}{c} \frac{l}{2} = \gamma_0 \frac{U_0}{2 E^{(0)}} \frac{c}{\Omega}. \tag{4.31}$$

In general, it can be found that $C_1 = 0$, this confirms the impossibility to excite the cosine eigenmodes. This fact is valid for all cases in this work. Thus, from now on all cosine eigenfunctions will be omitted.

The remaining constant C_2 can also be calculated, then the forced vibrations are obtained as

$$u(x,t) = \frac{\gamma_0 U_0 l}{4 E^{(0)} \lambda \cos \lambda} \sin\left(\frac{2\lambda}{l} x\right) \cos \Omega t, \tag{4.32}$$

with

$$\lambda = \Omega \frac{l}{2} \sqrt{\frac{\rho}{E^{(0)}}} \quad \text{and} \quad x \in \left[-\frac{l}{2}, \frac{l}{2}\right].$$

4.4 Ritz discretization

So far the field equation for the linear undamped system can be solved exactly, it is not necessary to use this approximation method with a spatial discretization. However, this discretization will be utilized later for the nonlinear case in chapter 5. Therefore, the discretization process should first be described here in detail for the simple case of a linear conservative system and then extended for the nonconservative case.

Using the Ritz ansatz [41]

$$u(x,t) = \sum_{k=1}^{n} U_k(x) \, p_k(t), \tag{4.33}$$

where $U_k(x)$ are the sinusoidal eigenfunctions of the longitudinal vibrations of piezoceramics found in section 4.3 and $p_k(t)$ are the unknown time functions which have to be determined, the variations of the displacement and its partial derivatives are

$$\delta u = \sum_{k=1}^{n} U_k(x) \, \delta p_k(t), \quad \delta \dot{u} = \sum_{k=1}^{n} U_k(x) \, \delta \dot{p}_k(t), \quad \delta u' = \sum_{k=1}^{n} U_k'(x) \, \delta p_k(t). \tag{4.34}$$

Chapter 4. Linear dynamic modeling

Applying Hamilton's principle (2.11) with the electric enthalpy density given in the equation (4.10) and the virtual work $\delta W = 0$ yields

$$A \int_{t_0}^{t_1} \int_{-\frac{l}{2}}^{\frac{l}{2}} \left(\rho \, \ddot{u} \delta u - E^{(0)} \, u' \delta u' + \gamma_0 \, E_z \, \delta u' \right) \mathrm{d}x \, \mathrm{d}t = 0, \qquad (4.35)$$

with the electric field given in (4.20), $E_z = \dfrac{U_0}{h} \cos \Omega t$.

Substituting the Ritz ansatz and the corresponding variations $\delta \dot{u}$ and $\delta u'$ into the above equation leads to

$$A \int_{t_0}^{t_1} \int_{-\frac{l}{2}}^{\frac{l}{2}} \left(\rho \sum_{i,k=1}^{n} U_i \, U_k \, \dot{p}_i \, \delta \dot{p}_k - E^{(0)} \sum_{i,k=1}^{n} U_i' \, U_k' \, p_i \, \delta p_k + \gamma_0 \, E_z \sum_{k=1}^{n} U_k' \, \delta p_k \right) \mathrm{d}x \, \mathrm{d}t = 0. \qquad (4.36)$$

The first summation can be integrated by parts with respect to time t as

$$\int_{t_0}^{t_1} \dot{p}_i \, \delta \dot{p}_k \, \mathrm{d}t = \left[\dot{p}_i \, \delta p_k \right]_{t_0}^{t_1} - \int_{t_0}^{t_1} \ddot{p}_i \, \delta p_k \, \mathrm{d}t = - \int_{t_0}^{t_1} \ddot{p}_i \, \delta p_k \, \mathrm{d}t \qquad (4.37)$$

since all variations δp_k vanish at t_0 and t_1. Therefore, the equation (4.36) can be rewritten as

$$A \int_{t_0}^{t_1} \sum_{k=1}^{n} \left[- \sum_{i=1}^{n} \ddot{p}_i \, \rho \int_{-\frac{l}{2}}^{\frac{l}{2}} U_i \, U_k \, \mathrm{d}x - \sum_{i=1}^{n} p_i \, E^{(0)} \int_{-\frac{l}{2}}^{\frac{l}{2}} U_i' \, U_k' \, \mathrm{d}x \right.$$
$$\left. + \gamma_0 \, E_z \int_{-\frac{l}{2}}^{\frac{l}{2}} U_k' \, \mathrm{d}x \right] \delta p_k \, \mathrm{d}t = 0 \quad (4.38)$$

Note that the eigenfunctions are orthogonal over the interval $-l/2 \leq x \leq l/2$ and the same for their derivatives with respect to x, namely

$$\int_{-\frac{l}{2}}^{\frac{l}{2}} U_i \, U_k \, \mathrm{d}x = 0 \quad \text{and} \quad \int_{-\frac{l}{2}}^{\frac{l}{2}} U_i' \, U_k' \, \mathrm{d}x = 0 \quad \text{if} \quad i \neq k \text{ and } i, k = 1, 2, 3, \ldots, n.$$

The equation (4.38) reduces to

$$A \int_{t_0}^{t_1} \sum_{k=1}^{n} \left[- \ddot{p}_k \, \rho \int_{-\frac{l}{2}}^{\frac{l}{2}} U_k^2 \, \mathrm{d}x - p_k \, E^{(0)} \int_{-\frac{l}{2}}^{\frac{l}{2}} U_k'^2 \, \mathrm{d}x \right.$$
$$\left. + \gamma_0 \, \frac{U_0}{h} \int_{-\frac{l}{2}}^{\frac{l}{2}} U_k' \, \mathrm{d}x \, \cos \Omega t \right] \delta p_k \, \mathrm{d}t = 0. \quad (4.39)$$

Since the variations δp_k are arbitrary, the expression in the square brackets in the equation (4.39) vanishes for all positive integers k. The equations of motion can then be derived as

$$m_k \, \ddot{p}_k + c_k^{(1)} \, p_k = f_k^{(1)} \cos \Omega t, \quad k = 1, 2, 3, \ldots, n \qquad (4.40)$$

with

$$m_k = \rho \int_{-\frac{l}{2}}^{\frac{l}{2}} U_k^2(x) \, \mathrm{d}x, \quad c_k^{(1)} = E^{(0)} \int_{-\frac{l}{2}}^{\frac{l}{2}} U_k'^2(x) \, \mathrm{d}x, \quad f_k^{(1)} = \gamma_0 \, \frac{U_0}{h} \int_{-\frac{l}{2}}^{\frac{l}{2}} U_k'(x) \, \mathrm{d}x.$$

4.5 Linear damping

Up to now the vibrating piezoceramic rods are only considered as conservative systems. But in fact dissipative forces always exist, thus it is necessary to introduce dissipation in the linear modeling of the system. According to [117], dissipative behavior can be described by a viscoelastic material constitutive relation and a piezoelectric as well as dielectric analog. So the linear constitutive equations (4.6) and (4.7) are extended with dissipative terms as

$$T_{xx} = E^{(0)} S_{xx} - \gamma_0 E_z + E_d^{(0)} \dot{S}_{xx} - \gamma_{0d} \dot{E}_z, \qquad (4.41)$$
$$D_z = \gamma_0 S_{xx} + \nu_0 E_z + \gamma_{0d} \dot{S}_{xx} + \nu_{0d} \dot{E}_z. \qquad (4.42)$$

Hence, the virtual work δW in Hamilton's principle is no longer equal to zero, namely

$$\delta W = -A \int_{-\frac{l}{2}}^{\frac{l}{2}} \left(E_d^{(0)} \dot{S}_{xx} - \gamma_{0d} \dot{E}_z \right) \delta u' \, dx. \qquad (4.43)$$

Substituting the electric enthalpy density H in (4.10) and the virtual work δW in (4.43) into Hamilton's principle (2.11) and considering Ritz discretization yields the linear equations of motion of the linear damped system as

$$m_k \ddot{p}_k + d_k \dot{p}_k + c_k^{(1)} p_k = f_k^{(1)} \cos \Omega t - f_{kd}^{(1)} \Omega \sin \Omega t, \quad k = 1, 2, 3, \ldots, n \qquad (4.44)$$

with

$$m_k = \rho \int_{-\frac{l}{2}}^{\frac{l}{2}} U_k^2(x) \, dx, \quad d_k = E_d^{(0)} \int_{-\frac{l}{2}}^{\frac{l}{2}} U_k'^2(x) \, dx, \quad c_k^{(1)} = E^{(0)} \int_{-\frac{l}{2}}^{\frac{l}{2}} U_k'^2(x) \, dx,$$

$$f_k^{(1)} = \gamma_0 \frac{U_0}{h} \int_{-\frac{l}{2}}^{\frac{l}{2}} U_k'(x) \, dx, \quad f_{kd}^{(1)} = \gamma_{0d} \frac{U_0}{h} \int_{-\frac{l}{2}}^{\frac{l}{2}} U_k'(x) \, dx.$$

Chapter 5

Nonlinear dynamic modeling

The nonlinear phenomena exhibited by piezoceramics observed so far can be classified according to the magnitude of excitations. For example, nonlinear phenomena with strong electrical loads are represented by hysteresis curves as shown in figures 2.4 and 2.5, where an occurrence of domain switching processes leads to the irreversible macroscopic polarization, this means the switched domain structure basically remains after the withdrawal of applied electric fields. This hysteretic behavior can be modeled e.g. by using a phenomenological consitutive modeling as well as a thermodynamically consistent constitutive one described in [63]. The polarization and strain are decomposed into reversible and irreversible parts. In contrast to the irreversible part, the reversible one vanishes after unloading. For the phenomenological modeling of dielectric and butterfly hysteresis presented in figures 2.4 and 2.5, respectively, the reversible part conforms to a linear constitutive law, whereas the irreversible one is represented for instance by piecewise linear evolution equations. In the thermodynamically based model, microscopically motivated internal variables are introduced. The macroscopic irreversible quantities are then assumed to be functions of these variables. The evolution equations for the internal variables are chosen in order to satisfy the Clausius-Duhem inequality.

By contrast, this chapter gives a dynamic modeling of nonlinear phenomena observed in experiments described in chapter 3 corresponding to weak electrical loads basically resulting in no switching process, i.e. the microscopic structure of the material and the remanent polarization remain unchanged. Based on the work of von Wagner [117] the nonlinear elastic, piezoelectric and dielectric terms are first considered in the electric enthalpy density. Then the linear and nonlinear dissipative terms are introduced into the nonlinear conservative constitutive equations. Using Hamilton's principle and a single-mode Ritz discretization gives rise to the nonlinear equation of motion. This equation can be approximately solved by using Lindstedt-Poincaré perturbation technique. Fitting the theoretical displacement amplitude responses to the experimental ones, parameters of piezoceramics can be determined. Along with an extra consideration for quadratic nonlinearities it can be seen, that the identification problem has multivalued solutions as already shown in [117]. This leads to an ambiguity in the decision on the type of nonlinearities to explain the observed nonlinear dynamic effects.

5.1 Nonlinear constitutive equations

The nonlinear dynamic behavior of tranversally polarized piezoceramics excited close to resonance by weak electric fields using the inverse 31-effect can be modeled based on adequate nonlinear constitutive relations. Due to the occurrence of both quadratic and cubic nonlinearities observed in experiments, the third and the fourth order terms are first introduced into the electric enthalpy density [90, 117]

$$\begin{aligned}
H = {} & \frac{1}{2} E^{(0)} S_{xx}^2 - \gamma_0 S_{xx} E_z - \frac{1}{2} \nu_0 E_z^2 \\
& + \frac{1}{3} E^{(1)} S_{xx}^3 - \frac{1}{2} \gamma_1^{(1)} S_{xx}^2 E_z - \frac{1}{2} \gamma_1^{(2)} S_{xx} E_z^2 - \frac{1}{3} \nu_1 E_z^3 \\
& + \frac{1}{4} E^{(2)} S_{xx}^4 - \frac{1}{3} \gamma_2^{(1)} S_{xx}^3 E_z - \frac{1}{2} \gamma_2^{(2)} S_{xx}^2 E_z^2 - \frac{1}{3} \gamma_2^{(3)} S_{xx} E_z^3 - \frac{1}{4} \nu_2 E_z^4,
\end{aligned} \quad (5.1)$$

where $E^{(1)}$ and $E^{(2)}$ are the parameters of the quadratic and cubic elastic terms respectively, $\gamma_1^{(1)}$, $\gamma_1^{(2)}$, $\gamma_2^{(1)}$, $\gamma_2^{(2)}$, $\gamma_2^{(3)}$ are the parameters of nonlinear piezoelectric coupling terms and ν_1, ν_2 are the parameters of the higher-order dielectric terms.

The nonlinear conservative constitutive equations result from (2.15) and afterwards are extended by both linear and nonlinear dissipative terms

$$\begin{aligned}
T_{xx} = {} & E^{(0)} S_{xx} - \gamma_0 E_z + E_d^{(0)} \dot{S}_{xx} - \gamma_{0d} \dot{E}_z \\
& + E^{(1)} S_{xx}^2 - \gamma_1^{(1)} S_{xx} E_z - \frac{1}{2} \gamma_1^{(2)} E_z^2 + E_d^{(1)} (\dot{S}_{xx}^2) - \gamma_{1d}^{(1)} (S_{xx} \dot{E}_z) - \frac{1}{2} \gamma_{1d}^{(2)} (\dot{E}_z^2) \\
& + E^{(2)} S_{xx}^3 - \gamma_2^{(1)} S_{xx}^2 E_z - \gamma_2^{(2)} S_{xx} E_z^2 - \frac{1}{3} \gamma_2^{(3)} E_z^3 \\
& + E_d^{(2)} (\dot{S}_{xx}^3) - \gamma_{2d}^{(1)} (S_{xx}^2 \dot{E}_z) - \gamma_{2d}^{(2)} (S_{xx} \dot{E}_z^2) - \frac{1}{3} \gamma_{2d}^{(3)} (\dot{E}_z^3),
\end{aligned} \quad (5.2)$$

$$\begin{aligned}
D_z = {} & \gamma_0 S_{xx} + \nu_0 E_z + \gamma_{0d} \dot{S}_{xx} + \nu_{0d} \dot{E}_z \\
& + \frac{1}{2} \gamma_1^{(1)} S_{xx}^2 + \gamma_1^{(2)} S_{xx} E_z + \nu_1 E_z^2 + \frac{1}{2} \gamma_{1d}^{(1)} (\dot{S}_{xx}^2) + \gamma_{1d}^{(2)} (S_{xx} \dot{E}_z) + \nu_{1d} (\dot{E}_z^2) \\
& + \frac{1}{3} \gamma_2^{(1)} S_{xx}^3 + \gamma_2^{(2)} S_{xx}^2 E_z + \gamma_2^{(3)} S_{xx} E_z^2 + \nu_2 E_z^3 \\
& + \frac{1}{3} \gamma_{2d}^{(1)} (\dot{S}_{xx}^3) + \gamma_{2d}^{(2)} (S_{xx}^2 \dot{E}_z) + \gamma_{2d}^{(3)} (S_{xx} \dot{E}_z^2) + \nu_{2d} (\dot{E}_z^3),
\end{aligned} \quad (5.3)$$

where $E_d^{(1)}$ and $E_d^{(2)}$ correspond to the parameters of the quadratic and cubic dissipative mechanical terms, $\gamma_{1d}^{(1)}$, $\gamma_{1d}^{(2)}$, $\gamma_{2d}^{(1)}$, $\gamma_{2d}^{(2)}$, $\gamma_{2d}^{(3)}$ are the parameters of nonlinear dissipative piezoelectric terms and ν_{1d}, ν_{2d} are the parameters of higher-order dissipative dielectric terms.

5.2 Ritz discretization

In order to obtain equations of motion, the nonlinear enthalpy density given in (5.1) and the virtual work formulated from the nonlinear constitutive equations (5.2) and (5.3) should be substituted into Hamilton's principle (2.11). However unlike the linear case, the electric field E_z in the nonlinear case is now coupled with the strain S_{xx} and can be derived as solution of Maxwell's equation

$$\frac{\partial D_z}{\partial z} = 0. \qquad (5.4)$$

Applying perturbation technique with the damping terms and the nonlinear terms considered to be of order ε this nonlinear partial differential equation can be solved for the zeroth approximation of E_z which is the same as that of the linear case in (4.20)

$$E_z = \frac{U_0}{h} \cos \Omega t. \qquad (5.5)$$

Furthermore in the resonance case, the weak excitation electric field is considered to be of order ε, thus the higher order approximations of the electric field have no influence on the zeroth approximation of the displacement. Therefore, the eletric field can also be assumed to be homogeneous and given in the equation (5.5) for the nonlinear case [117]. This means the small variation of the electric field δE_z vanishes.

For the discretization, the Ritz ansatz (4.33) is used again. The shape functions $U_k(x)$ are still the sinusoidal eigenfunctions (4.24) of the linear system. In the linear case, owing to the orthogonality of the shape functions, the discretized equations of motion are decoupled. By contrast, due to the nonlinear terms the separate differential equations can no longer be generated. For this reason, the Ritz ansatz (4.33) is restricted to a single eigenfunction, which corresponds to the eigenfrequency close to the excitation frequency (the first resonance frequency in this work). It has been verified that using only one eigenfunction gives sufficiently accurate results [90, 118, 120].

Considering the excitation near to the k-th eigenfrequency, the single-mode Ritz ansatz is used as

$$u(x,t) = U_k(x) \, p_k(t). \qquad (5.6)$$

Substituting this into Hamilton's principle and performing the respective variations yields the nonlinear equation of motion for the longitudinal vibrations of the piezoceramic rod close to the k-th resonance

$$\begin{aligned}
& m_k \ddot{p}_k + d_k \dot{p}_k + c_k^{(1)} p_k + c_k^{(2)} p_k^2 + c_{kd}^{(2)} p_k \dot{p}_k + c_k^{(3)} p_k^3 + c_{kd}^{(3)} p_k^2 \dot{p}_k \\
&= f_k^{(1)} \cos \Omega t - f_{kd}^{(1)} \Omega \sin \Omega t + f_k^{(2)} p_k \cos \Omega t + f_{kd}^{(2)} \dot{p}_k \cos \Omega t - f_{kd}^{(2)} p_k \Omega \sin \Omega t \\
&\quad + f_k^{(3)} \cos^2 \Omega t - f_{kd}^{(3)} \Omega \cos \Omega t \sin \Omega t + f_k^{(4)} p_k^2 \cos \Omega t + 2 f_{kd}^{(4)} p_k \dot{p}_k \cos \Omega t \\
&\quad - f_{kd}^{(4)} p_k^2 \Omega \sin \Omega t + f_k^{(5)} p_k \cos^2 \Omega t + f_{kd}^{(5)} \dot{p}_k \cos^2 \Omega t \\
&\quad - 2 f_{kd}^{(5)} p_k \Omega \cos \Omega t \sin \Omega t + f_k^{(6)} \cos^3 \Omega t - f_{kd}^{(6)} \Omega \cos^2 \Omega t \sin \Omega t, \qquad (5.7)
\end{aligned}$$

with

$$m_k = \rho \int_{-\frac{l}{2}}^{\frac{l}{2}} U_k^2(x)\,dx, \quad d_k = E_d^{(0)} \int_{-\frac{l}{2}}^{\frac{l}{2}} U_k'^2(x)\,dx, \quad c_k^{(1)} = E^{(0)} \int_{-\frac{l}{2}}^{\frac{l}{2}} U_k'^2(x)\,dx,$$

$$c_k^{(2)} = E^{(1)} \int_{-\frac{l}{2}}^{\frac{l}{2}} U_k'^3(x)\,dx, \quad c_{kd}^{(2)} = 2 E_d^{(1)} \int_{-\frac{l}{2}}^{\frac{l}{2}} U_k'^3(x)\,dx,$$

$$c_k^{(3)} = E^{(2)} \int_{-\frac{l}{2}}^{\frac{l}{2}} U_k'^4(x)\,dx, \quad c_{kd}^{(3)} = 3 E_d^{(2)} \int_{-\frac{l}{2}}^{\frac{l}{2}} U_k'^4(x)\,dx,$$

$$f_k^{(1)} = \gamma_0 \frac{U_0}{h} \int_{-\frac{l}{2}}^{\frac{l}{2}} U_k'(x)\,dx, \quad f_{kd}^{(1)} = \gamma_{0d} \frac{U_0}{h} \int_{-\frac{l}{2}}^{\frac{l}{2}} U_k'(x)\,dx,$$

$$f_k^{(2)} = \gamma_1^{(1)} \frac{U_0}{h} \int_{-\frac{l}{2}}^{\frac{l}{2}} U_k'^2(x)\,dx, \quad f_{kd}^{(2)} = \gamma_{1d}^{(1)} \frac{U_0}{h} \int_{-\frac{l}{2}}^{\frac{l}{2}} U_k'^2(x)\,dx,$$

$$f_k^{(3)} = \frac{1}{2} \gamma_1^{(2)} \frac{U_0^2}{h^2} \int_{-\frac{l}{2}}^{\frac{l}{2}} U_k'(x)\,dx, \quad f_{kd}^{(3)} = \gamma_{1d}^{(2)} \frac{U_0^2}{h^2} \int_{-\frac{l}{2}}^{\frac{l}{2}} U_k'(x)\,dx,$$

$$f_k^{(4)} = \gamma_2^{(1)} \frac{U_0}{h} \int_{-\frac{l}{2}}^{\frac{l}{2}} U_k'^3(x)\,dx, \quad f_{kd}^{(4)} = \gamma_{2d}^{(1)} \frac{U_0}{h} \int_{-\frac{l}{2}}^{\frac{l}{2}} U_k'^3(x)\,dx,$$

$$f_k^{(5)} = \gamma_2^{(2)} \frac{U_0^2}{h^2} \int_{-\frac{l}{2}}^{\frac{l}{2}} U_k'^2(x)\,dx, \quad f_{kd}^{(5)} = \gamma_{2d}^{(2)} \frac{U_0^2}{h^2} \int_{-\frac{l}{2}}^{\frac{l}{2}} U_k'^2(x)\,dx,$$

$$f_k^{(6)} = \frac{1}{3} \gamma_2^{(3)} \frac{U_0^3}{h^3} \int_{-\frac{l}{2}}^{\frac{l}{2}} U_k'(x)\,dx, \quad f_{kd}^{(6)} = \gamma_{2d}^{(3)} \frac{U_0^3}{h^3} \int_{-\frac{l}{2}}^{\frac{l}{2}} U_k'(x)\,dx.$$

5.3 Solution by perturbation analysis

In order to find out an approximation of the periodic solution of the equation of motion (5.7), the respective dimensionless equation will be attacked using perturbation technique [40, 87]. First the nondimensional time τ and the ratio η of the excitation frequency to the natural frequency are introduced [117]

$$\tau = \omega_0 t \quad \text{with} \quad \omega_0 = \sqrt{c_k^{(1)}/m_k}, \tag{5.8}$$

$$\eta = \frac{\Omega}{\omega_0}. \tag{5.9}$$

Considering the phase shift between the excitation voltage and the displacement response, the phase of the excitation is set as $(\Omega t + \psi)$ instead of Ωt and that in nondimensional form corresponds to

$$\cos \eta\tau \;\to\; \cos(\eta\tau + \psi) \;=\; h_1 \cos \eta\tau + h_2 \sin \eta\tau, \tag{5.10}$$
$$\sin \eta\tau \;\to\; \sin(\eta\tau + \psi) \;=\; -h_2 \cos \eta\tau + h_1 \sin \eta\tau, \tag{5.11}$$

where $h_1 = \cos \psi$ and $h_2 = -\sin \psi$, thus

$$h_1^2 + h_2^2 = 1. \tag{5.12}$$

Chapter 5. Nonlinear dynamic modeling

Introducing the "small" parameter ε and assuming that the quadratic and cubic nonlinearities, damping terms as well as linear excitation terms are of the first-order ε [117], the discretized equation of motion (5.7) can be expressed as

$$p'' + \varepsilon k p' + (1^2) p + \varepsilon \alpha_1^* p^2 + \varepsilon \alpha_{1d}^* p p' + \varepsilon \alpha_1 p^3 + \varepsilon \alpha_{1d} p^2 p'$$
$$= \varepsilon q (h_1 \cos \eta \tau + h_2 \sin \eta \tau) + \varepsilon q_d \eta (-h_2 \cos \eta \tau + h_1 \sin \eta \tau)$$
$$+ \varepsilon \alpha_2^* p (h_1 \cos \eta \tau + h_2 \sin \eta \tau) + \varepsilon \alpha_{2d}^* p' (h_1 \cos \eta \tau + h_2 \sin \eta \tau)$$
$$- \varepsilon \alpha_{2d}^* p \eta (-h_2 \cos \eta \tau + h_1 \sin \eta \tau) + \varepsilon^2 \alpha_3^* (h_1 \cos \eta \tau + h_2 \sin \eta \tau)^2$$
$$- \varepsilon^2 \alpha_{3d}^* \eta (h_1 \cos \eta \tau + h_2 \sin \eta \tau)(-h_2 \cos \eta \tau + h_1 \sin \eta \tau)$$
$$+ \varepsilon \alpha_2 p^2 (h_1 \cos \eta \tau + h_2 \sin \eta \tau) + 2 \varepsilon \alpha_{2d} p p' (h_1 \cos \eta \tau + h_2 \sin \eta \tau)$$
$$- \varepsilon \alpha_{2d} p^2 \eta (-h_2 \cos \eta \tau + h_1 \sin \eta \tau)$$
$$+ \varepsilon^2 \alpha_3 p (h_1 \cos \eta \tau + h_2 \sin \eta \tau)^2 + \varepsilon^2 \alpha_{3d} p' (h_1 \cos \eta \tau + h_2 \sin \eta \tau)^2$$
$$- 2 \varepsilon^2 \alpha_{3d} p \eta (h_1 \cos \eta \tau + h_2 \sin \eta \tau)(-h_2 \cos \eta \tau + h_1 \sin \eta \tau)$$
$$+ \varepsilon^3 \alpha_4 (h_1 \cos \eta \tau + h_2 \sin \eta \tau)^3$$
$$- \varepsilon^3 \alpha_{4d} \eta (h_1 \cos \eta \tau + h_2 \sin \eta \tau)^2 (-h_2 \cos \eta \tau + h_1 \sin \eta \tau), \qquad (5.13)$$

where

$$()' = \frac{\mathrm{d}}{\mathrm{d}\tau}, \quad k = \frac{d}{\varepsilon m_k \omega_0}, \quad \alpha_1^* = \frac{c_k^{(2)}}{\varepsilon m_k \omega_0^2}, \quad \alpha_{1d}^* = \frac{c_{kd}^{(2)}}{\varepsilon m_k \omega_0},$$

$$\alpha_1 = \frac{c_k^{(3)}}{\varepsilon m_k \omega_0^2}, \quad \alpha_{1d} = \frac{c_{kd}^{(3)}}{\varepsilon m_k \omega_0}, \quad q = \frac{f_k^{(1)}}{\varepsilon m_k \omega_0^2}, \quad q_d = -\frac{f_{kd}^{(1)}}{\varepsilon m_k \omega_0},$$

$$\alpha_2^* = \frac{f_k^{(2)}}{\varepsilon m_k \omega_0^2}, \quad \alpha_{2d}^* = \frac{f_{kd}^{(2)}}{\varepsilon m_k \omega_0}, \quad \alpha_3^* = \frac{f_k^{(3)}}{\varepsilon^2 m_k \omega_0^2}, \quad \alpha_{3d}^* = \frac{f_{kd}^{(3)}}{\varepsilon^2 m_k \omega_0},$$

$$\alpha_2 = \frac{f_k^{(4)}}{\varepsilon m_k \omega_0^2}, \quad \alpha_{2d} = \frac{f_{kd}^{(4)}}{\varepsilon m_k \omega_0}, \quad \alpha_3 = \frac{f_k^{(5)}}{\varepsilon^2 m_k \omega_0^2}, \quad \alpha_{3d} = \frac{f_{kd}^{(5)}}{\varepsilon^2 m_k \omega_0},$$

$$\alpha_4 = \frac{f_k^{(6)}}{\varepsilon^3 m_k \omega_0^2}, \quad \alpha_{4d} = \frac{f_{kd}^{(6)}}{\varepsilon^3 m_k \omega_0}$$

and the subscript k is dropped for simplicity of the notation.

Applying the Lindstedt-Poincaré method considering the excitation near to the natural frequency, an expansion for η is defined as

$$\eta = 1 + \varepsilon \eta_1 + \varepsilon^2 \eta_2 + \cdots. \qquad (5.14)$$

Hence,

$$1^2 = \eta^2 - 2\varepsilon \eta_1 + \cdots. \qquad (5.15)$$

Assuming that $p(\tau)$ can be found as an expansion of the following form

$$p(\tau) = p_0(\tau) + \varepsilon p_1(\tau) + \varepsilon^2 p_2(\tau) + \cdots, \qquad (5.16)$$

substituting (5.15) and (5.16) into the equation (5.13), then equating the coefficients of ε^0 and ε^1 on both sides yields

$$\varepsilon^0: \quad p_0'' + \eta^2 p_0 = 0, \tag{5.17}$$

$$\begin{aligned}\varepsilon^1: \quad p_1'' + \eta^2 p_1 =& -k p_0' + 2\eta_1 p_0 - \alpha_1^* p_0^2 - \alpha_{1d}^* p_0 p_0' - \alpha_1 p_0^3 \\ & - \alpha_{1d} p_0^2 p_0' + q\left(h_1 \cos \eta\tau + h_2 \sin \eta\tau\right) + q_d \eta\left(-h_2 \cos \eta\tau + h_1 \sin \eta\tau\right) \\ & + \alpha_2^* p_0 \left(h_1 \cos \eta\tau + h_2 \sin \eta\tau\right) + \alpha_{2d}^* p_0' \left(h_1 \cos \eta\tau + h_2 \sin \eta\tau\right) \\ & - \alpha_{2d}^* p_0 \eta\left(-h_2 \cos \eta\tau + h_1 \sin \eta\tau\right) + \alpha_2 p_0^2 \left(h_1 \cos \eta\tau + h_2 \sin \eta\tau\right) \\ & + 2\alpha_{2d} p_0 p_0' \left(h_1 \cos \eta\tau + h_2 \sin \eta\tau\right) - \alpha_{2d} p_0^2 \eta\left(-h_2 \cos \eta\tau + h_1 \sin \eta\tau\right). \end{aligned} \tag{5.18}$$

Since the phase shift ψ between the excitation voltage and the displacement response has already been taken into account, the solution of the equation (5.17) is given as

$$p_0(\tau) = A \cos \eta\tau, \tag{5.19}$$

where the amplitude A is unknown. Substituting the zeroth approximation (5.19) into the equation (5.18) and eliminating the secular terms [40, 87] yields

$$2\eta_1 A - \frac{3}{4}\alpha_1 A^3 + \left(q + \frac{3}{4}\alpha_2 A^2\right) h_1 + \left(-q_d + \frac{1}{4}\alpha_{2d} A^2\right) \eta h_2 = 0, \tag{5.20}$$

$$k\eta A + \frac{1}{4}\alpha_{1d}\eta A^3 + \left(q_d - \frac{3}{4}\alpha_{2d} A^2\right) \eta h_1 + \left(q + \frac{1}{4}\alpha_2 A^2\right) h_2 = 0, \tag{5.21}$$

with three unknown variables A, h_1 and h_2. Solving these equations for h_1, h_2 and introducing them into the equation (5.12) leads to a fifth-order polynomial equation with respect to A^2, from which the stationary amplitude A of the zeroth approximation response can be found. Then the phase shift ψ can be determined as

$$\arctan \psi = -\frac{h_2}{h_1}. \tag{5.22}$$

It can be observed that the quadratic parameters $E^{(1)}$ and $E_d^{(1)}$, the nonlinear piezoelectric parameters $\gamma_1^{(1)}$, $\gamma_{1d}^{(1)}$, $\gamma_1^{(2)}$, $\gamma_{1d}^{(2)}$, $\gamma_2^{(2)}$, $\gamma_{2d}^{(2)}$, $\gamma_2^{(3)}$ and $\gamma_{2d}^{(3)}$ have no influence on the solution of the perturbation method. In addition, due to the assumption of the given excitation electric field E_z, all dielectric parameters ν_0, ν_{0d}, ν_1, ν_{1d}, ν_2, ν_{2d} are also absent from the solution. However it can later be seen in section 5.5, that the quadratic terms will affect the zeroth approximation of the solution if they are taken of order ε while the cubic terms, damping terms and the linear excitation terms are of order ε^2. Although this trick is not verified by experiments, where superharmonics at the first resonance frequency of comparable order of magnitude can be observed by excitation at one half or one third of the first resonance frequency [90, 117], quadratic nonlinear behavior is always expected to occur with piezoceramics which are structured by asymmetric unit cells.

Chapter 5. Nonlinear dynamic modeling

5.4 Determination of parameters

In this section, the parameter identification for the piezoceramic materials is presented. The parameters will be determined in $Mathematica$, such that the deviation between the theoretical displacement amplitude–frequency curves near the first resonance and the respective experimental ones is minimized by using the built-in function $FindFit$ which finds a least-squares fit. Moreover, these parameters can manually be corrected with the help of a $Mathematica$ program written by the author himself.

It has been pointed out in section 5.3 that a series of nonlinear parameters does not take part in the zeroth approximation response of the piezoceramic rods. Therefore, only the following parameters can be determined by fitting in the modeling and measuring results

- Linear elastic modulus $E^{(0)}$ and linear piezoelectric parameter γ_0,
- Parameters of linear damping terms $E_d^{(0)}$ and γ_{0d},
- Parameters of cubic conservative terms $E^{(2)}$ and $\gamma_2^{(1)}$,
- Parameters of cubic dissipative terms $E_d^{(2)}$ and $\gamma_{2d}^{(1)}$.

In order to utilize the identification program, the influence of the material parameters on the displacement amplitude response should be considered, for example in the neighborhood of a response curve corresponding to a set of chosen parameters. The following is such consideration for the case of piezoceramic rods of PIC 181 with the dimensions $30 \times 3 \times 2$ mm^3 excited close to the first resonance frequency at the voltage amplitude of 15 V.

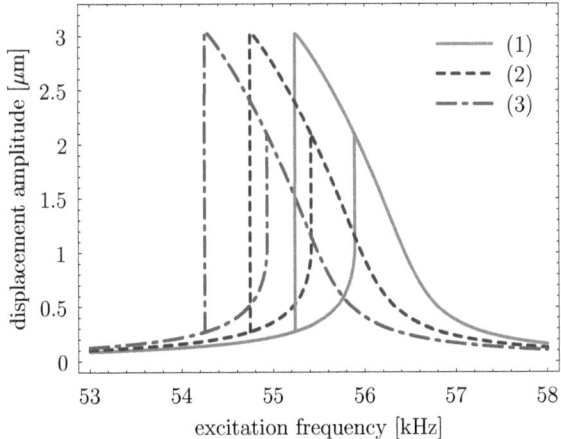

Figure 5.1: Theoretical displacement amplitude responses of PIC 181 at 15 V for different $E^{(0)} \left[\dfrac{\text{N}}{\text{m}^2}\right]$: $E^{(0)} = 8.95 \times 10^{10}$ (1), $E^{(0)} = 8.80 \times 10^{10}$ (2), $E^{(0)} = 8.65 \times 10^{10}$ (3).

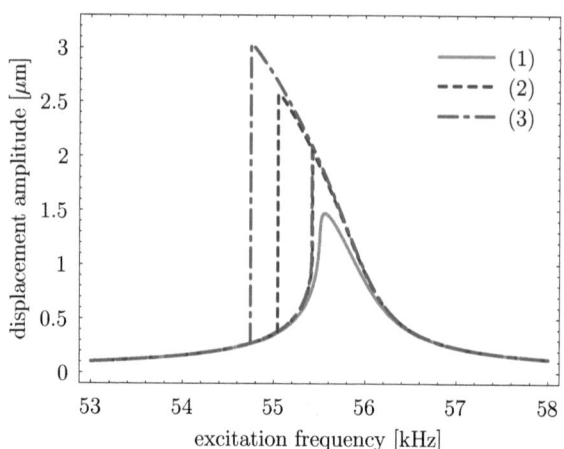

Figure 5.2: Theoretical displacement amplitude responses of PIC 181 at 15 V for different $E_d^{(0)}\left[\dfrac{\text{Ns}}{\text{m}^2}\right]$: $E_d^{(0)}=1600$ (1), $E_d^{(0)}=600$ (2), $E_d^{(0)}=300$ (3).

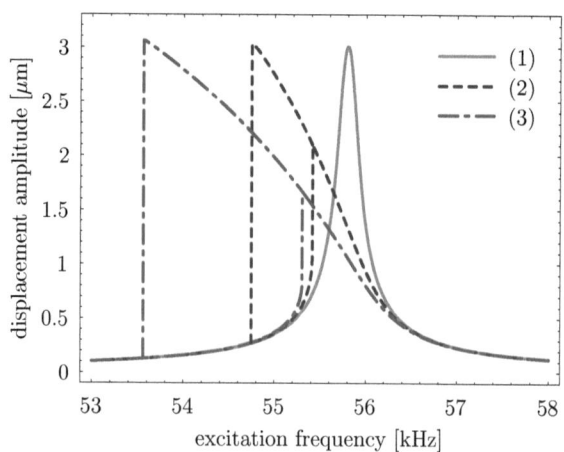

Figure 5.3: Theoretical displacement amplitude responses of PIC 181 at 15 V for different $E^{(2)}\left[\dfrac{\text{N}}{\text{m}^2}\right]$: $E^{(2)}=0$ (1), $E^{(2)}=-5.8\times 10^{16}$ (2), $E^{(2)}=-12.0\times 10^{16}$ (3).

Chapter 5. Nonlinear dynamic modeling

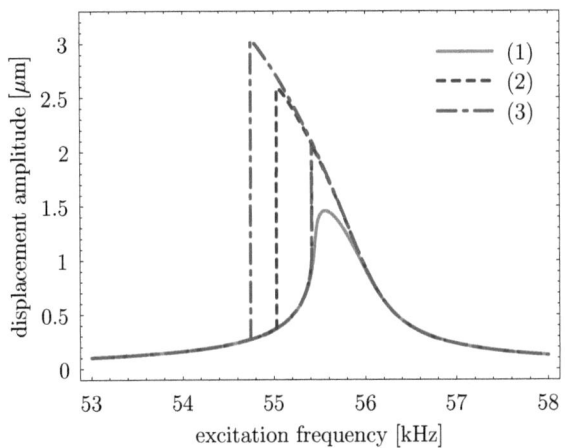

Figure 5.4: Theoretical displacement amplitude responses of PIC 181 at 15 V for different $E_d^{(2)} \left[\dfrac{\text{Ns}}{\text{m}^2}\right]$: $E_d^{(2)} = 110 \times 10^9$ (1), $E_d^{(2)} = 16 \times 10^9$ (2), $E_d^{(2)} = 9.05 \times 10^9$ (3).

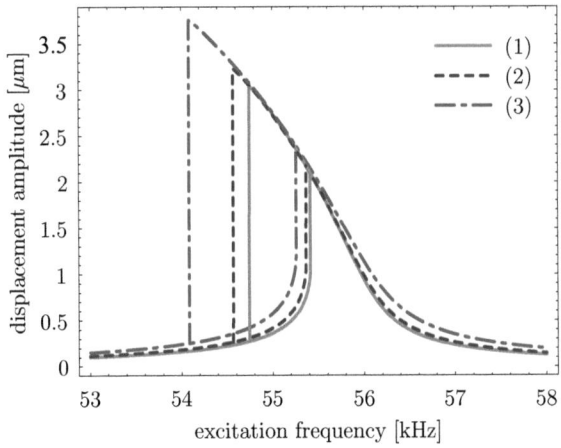

Figure 5.5: Theoretical displacement amplitude responses of PIC 181 at 15 V for different $\gamma_0 \left[\dfrac{\text{N}}{\text{Vm}}\right]$: $\gamma_0 = -4.5$ (1), $\gamma_0 = -7.0$ (2), $\gamma_0 = -9.8$ (3).

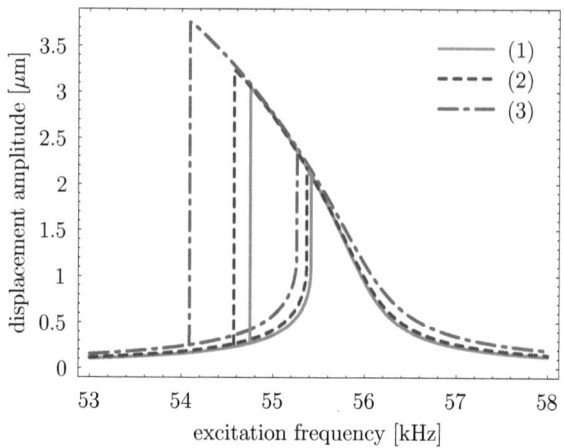

Figure 5.6: Theoretical displacement amplitude responses of PIC 181 at 15 V for different $\gamma_{0d} \left[\dfrac{\text{Ns}}{\text{Vm}}\right]$: $\gamma_{0d} = 0$ (1), $\gamma_{0d} = 1.0 \times 10^{-5}$ (2), $\gamma_{0d} = 2.0 \times 10^{-5}$ (3).

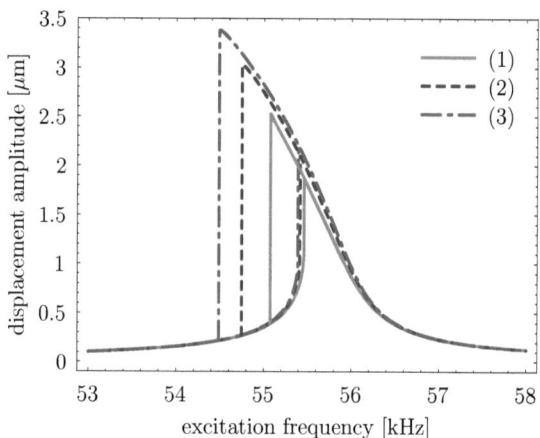

Figure 5.7: Theoretical displacement amplitude responses of PIC 181 at 15 V for different $\gamma_2^{(1)} \left[\dfrac{\text{N}}{\text{Vm}}\right]$: $\gamma_2^{(1)} = 3.0 \times 10^8$ (1), $\gamma_2^{(1)} = 0.3 \times 10^8$ (2), $\gamma_2^{(1)} = -1.0 \times 10^8$ (3).

Chapter 5. Nonlinear dynamic modeling

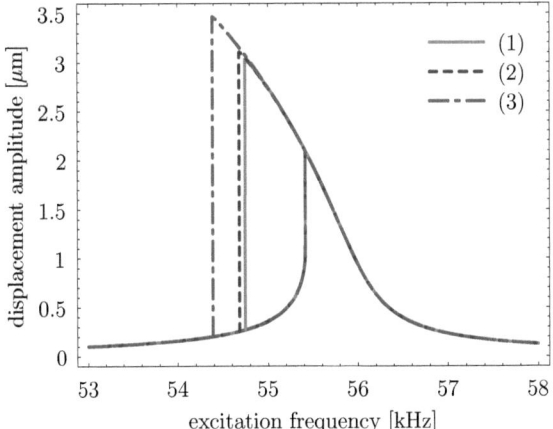

Figure 5.8: Theoretical displacement amplitude responses of PIC 181 at 15 V for different $\gamma_{2d}^{(1)} \left[\frac{\text{Ns}}{\text{Vm}}\right]$: $\gamma_{2d}^{(1)} = 0$ (1), $\gamma_{2d}^{(1)} = 200$ (2), $\gamma_{2d}^{(1)} = 400$ (3).

In each of figures 5.1–5.8 a parameter varies, whereas the remaining parameters take the values in the following set

$$E^{(0)} = 8.8 \times 10^{10} \frac{\text{N}}{\text{m}^2}, \quad E_d^{(0)} = 300 \frac{\text{Ns}}{\text{m}^2}, \quad E^{(2)} = -5.8 \times 10^{16} \frac{\text{N}}{\text{m}^2},$$

$$E_d^{(2)} = 9.05 \times 10^9 \frac{\text{Ns}}{\text{m}^2}, \quad \gamma_0 = -9.8 \frac{\text{N}}{\text{Vm}}, \quad \gamma_2^{(1)} = 0.3 \times 10^8 \frac{\text{N}}{\text{Vm}}, \quad \gamma_{0d} = \gamma_{2d}^{(1)} = 0 \frac{\text{Ns}}{\text{Vm}}.$$

From these figures it can be inferred how the resonance curve changes according to a "small" change in the parameters.

- $E^{(0)}$: Figure 5.1 shows that decreasing the linear elastic modulus will move the displacement amplitude response curve to the left. This means the first resonance frequency decreases, but the resonance amplitude and the difference between the two jump frequencies nearly remains.

- $E_d^{(0)}$: It can be seen in figure 5.2, that jump phenomena can not be observed with the linear damping of sufficient large magnitude. If the linear damping is decreased, under the influence of the cubic nonlinearities the jump phenomena may occur and the range of multiple solutions is greater.

- $E^{(2)}$: As shown in figure 5.3, the resonance curves bend more to the left when the negative cubic elastic parameter increases in absolute value. Thus the range of multiple stable responses is extended. Like in the case of $E^{(0)}$, the maximal amplitude almost remains.

- $E_d^{(2)}$: The influence of the cubic damping shown in figure 5.4 is basically similar to that of the linear damping.

- γ_0: Figure 5.5 presents an increase in response amplitude with increasing the linear piezoelectric parameter in absolute value. The jump frequencies decrease as a result and the range of multiple solutions is also greater. That is because the amplitude $f_k^{(1)}$ of the harmonic excitation in the equation of motion (5.7) is directly proportional to the parameter γ_0.

- γ_{0d}: It is demonstrated in figure 5.6 that if the linear dissipative piezoelectric parameter increases, then the resonance curves are enlarged. This is not consistent with the condition that the corresponding term has to only dissipate energy of the system. Moreover, this is also valid in case $\gamma_{0d} < 0$.

- $\gamma_2^{(1)}$: In figure 5.7 decreasing the cubic conservative piezoelectric parameter enlarges the displacement amplitude responses even if this parameter is negative. In contrast to the case of the damping parameters $E_d^{(0)}$ and $E_d^{(2)}$, the jump phenomena always occur.

- $\gamma_{2d}^{(1)}$: Figure 5.8 indicates that the cubic dissipative piezoelectric parameter has a small influence on the resonance curves and that is similar to in the case of γ_{0d}. Extension of the resonance curves with increasing $\gamma_{2d}^{(1)}$ is also not appropriate to the dissipative prerequisite for this parameter.

In all cases except for the elastic modulus $E^{(0)}$, only small changes in the jump-up frequencies can be observed. Due to the inconsistency of the dissipative piezoelectric parameters γ_{0d} and $\gamma_{2d}^{(1)}$, they will be set equal to zero in the present work.

5.4.1 Parameter identification from linear behavior

In the first step, the piezoceramic samples are excited close to the first resonance by low excitation voltages, so that an approximately linear behavior can be observed. By fitting the experimental displacement resonance curves, the linear parameters $E^{(0)}$, γ_0 and $E_d^{(0)}$ can be found, if all nonlinear parameters are put to be equal to zero. In other words, from the equation (4.8) the elastic constant s_{11}^E, the piezoelectric constant d_{31} and the linear damping parameter $E_d^{(0)}$ are obtained. Here the mass density ρ takes the values of the manufacturer PI Ceramic.

For the piezoceramic PIC 255, the linear parameters given by the manufacturer are

$$\rho = 7800 \; \frac{\text{kg}}{\text{m}^3}, \quad s_{11}^E = 1.59 \times 10^{-11} \; \frac{\text{m}^2}{\text{N}}, \quad d_{31} = -1.74 \times 10^{-10} \; \frac{\text{m}}{\text{V}}.$$

$s_{11}^E \; \left[\dfrac{\text{m}^2}{\text{N}}\right]$	$d_{31} \; \left[\dfrac{\text{m}}{\text{V}}\right]$	$E_d^{(0)} \; \left[\dfrac{\text{Ns}}{\text{m}^2}\right]$	$\gamma_{0d} \; \left[\dfrac{\text{Ns}}{\text{Vm}}\right]$
1.662×10^{-11}	-1.88×10^{-10}	6035	0

Table 5.1: Identified linear parameters for PIC 255.

Using the optimized linear parameters given in table 5.1 for the excitation voltage of 2 V, a very good coincidence between theoretical result and measurement can be derived as shown in figure 5.9.

Chapter 5. Nonlinear dynamic modeling

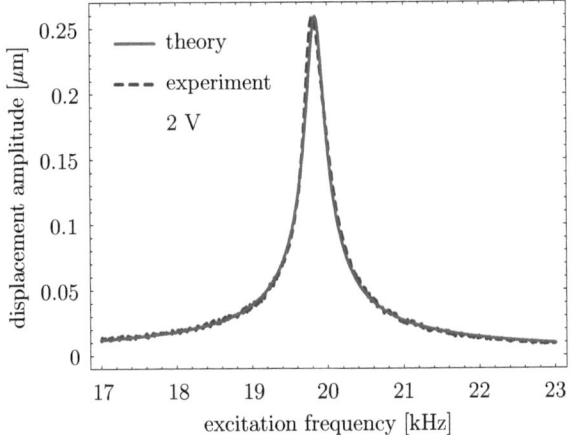

Figure 5.9: Linear parameters optimization for PIC 255.

For the piezoceramic PIC 181, the linear parameters given by the manufacturer are

$$\rho = 7850 \ \frac{\text{kg}}{\text{m}^3}, \quad s_{11}^E = 1.175 \times 10^{-11} \ \frac{\text{m}^2}{\text{N}}, \quad d_{31} = -1.08 \times 10^{-10} \ \frac{\text{m}}{\text{V}}.$$

The optimized linear parameters are given in table 5.2. A very good coincidence between theoretical and experimental results with respect to the excitation voltage of 0.5 V can also be seen in figure 5.10. It is obvious that the determined values for s_{11}^E and d_{31} are in good agreement with those of the manufacturer.

$s_{11}^E \ \left[\frac{\text{m}^2}{\text{N}}\right]$	$d_{31} \ \left[\frac{\text{m}}{\text{V}}\right]$	$E_d^{(0)} \ \left[\frac{\text{Ns}}{\text{m}^2}\right]$	$\gamma_{0d} \ \left[\frac{\text{Ns}}{\text{Vm}}\right]$
1.136×10^{-11}	-1.11×10^{-10}	285.91	0

Table 5.2: Identified linear parameters for PIC 181.

In addition, the linear dielectric parameters ν_0 and ν_{0d} in principle can be determined from the electric current amplitude responses of piezoceramics excited close to resonance. In this case the electric current through the piezoceramics can be calculated as [90, 117]

$$I(t) = \frac{\mathrm{d}Q}{\mathrm{d}t} = -\frac{\mathrm{d}}{\mathrm{d}t} \int_F D_z(x,t) \, \mathrm{d}F, \tag{5.23}$$

where Q is the electric charge transferred through one electrode and F is the area of the electrode. The electric displacement density is given in the linear constitutive equation (4.42) and with $\gamma_{0d} = 0$ this reduces to

$$D_z = \gamma_0 \, S_{xx} + \nu_0 \, E_z + \nu_{0d} \, \dot{E}_z, \tag{5.24}$$

where the linear electric field given in (4.20) is used taking account of the coefficients h_1 and h_2 which represents the phase shift between the displacement response and the

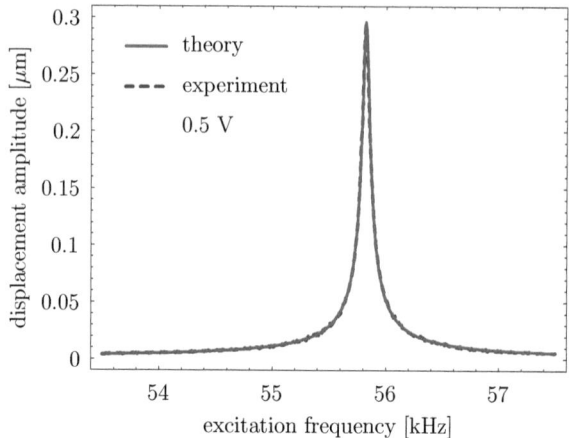

Figure 5.10: Linear parameters optimization for PIC 181.

excitation voltage. Using the strain–displacement relation (2.5), $S_{xx} = u'$, and the displacement for the first resonance case in the form

$$u(x,t) = A \sin \frac{\pi x}{l} \cos \Omega t, \qquad (5.25)$$

the electric current can be found from the equations (5.23) and (5.24) as

$$I(t) = -b \frac{d}{dt} \int_{-\frac{l}{2}}^{\frac{l}{2}} \left[\gamma_0 A \frac{\pi}{l} \cos \frac{\pi x}{l} \cos \Omega t + \nu_0 \frac{U_0}{h} (h_1 \cos \Omega t + h_2 \sin \Omega t) \right.$$
$$\left. + \nu_{0d} \frac{U_0}{h} \Omega (h_2 \cos \Omega t - h_1 \sin \Omega t) \right] dx, \qquad (5.26)$$

where l and b correspond to the length and the width of the electrode as well as of the piezoceramic samples. From the equation (5.26) the electric current amplitude can be calculated as

$$I_0 = b\Omega \sqrt{\left(2\gamma_0 A + \nu_0 l \frac{U_0}{h} h_1 + \nu_{0d} \Omega h_2\right)^2 + \left(\nu_{0d} \Omega h_1 - \nu_0 l \frac{U_0}{h} h_2\right)^2}. \qquad (5.27)$$

Fitting this electric current amplitude response to the experimental results, the parameters ν_0 and ν_{0d} can be obtained.

5.4.2 Parameter identification from nonlinear behavior

In order to identify the nonlinear parameters of piezoceramics, the samples are again excited near to the first resonance but by higher excitation voltages. The linear conservative and nonlinear parameters can be derived by fitting the theoretical displacement responses with the experimental curves, where the linear damping $E_d^{(0)}$ obtained in section 5.4.1 is used in identification process.

Chapter 5. Nonlinear dynamic modeling

U_0 [V]	s_{11}^E $\left[\frac{\text{m}^2}{\text{N}}\right]$	d_{31} $\left[\frac{\text{m}}{\text{V}}\right]$	$E_d^{(0)}$ $\left[\frac{\text{Ns}}{\text{m}^2}\right]$	$E^{(2)}$ $\left[\frac{\text{N}}{\text{m}^2}\right]$	$E_d^{(2)}$ $\left[\frac{\text{Ns}}{\text{m}^2}\right]$	γ_{0d} $\left[\frac{\text{Ns}}{\text{Vm}}\right]$	$\gamma_2^{(1)}$ $\left[\frac{\text{N}}{\text{Vm}}\right]$	$\gamma_{2d}^{(1)}$ $\left[\frac{\text{Ns}}{\text{Vm}}\right]$
20	1.679×10^{-11}	-1.90×10^{-10}	6035	-1.08×10^{18}	4.40×10^{12}	0	1.87×10^{8}	0
30	1.690×10^{-11}	-1.91×10^{-10}	6035	-8.02×10^{17}	3.58×10^{12}	0	1.87×10^{8}	0
40	1.697×10^{-11}	-1.92×10^{-10}	6035	-6.73×10^{17}	2.92×10^{12}	0	1.87×10^{8}	0
50	1.706×10^{-11}	-1.93×10^{-10}	6035	-6.24×10^{17}	2.72×10^{12}	0	1.87×10^{8}	0

Table 5.3: Identified parameters for PIC 255.

U_0 [V]	s_{11}^E $\left[\frac{\text{m}^2}{\text{N}}\right]$	d_{31} $\left[\frac{\text{m}}{\text{V}}\right]$	$E_d^{(0)}$ $\left[\frac{\text{Ns}}{\text{m}^2}\right]$	$E^{(2)}$ $\left[\frac{\text{N}}{\text{m}^2}\right]$	$E_d^{(2)}$ $\left[\frac{\text{Ns}}{\text{m}^2}\right]$	γ_{0d} $\left[\frac{\text{Ns}}{\text{Vm}}\right]$	$\gamma_2^{(1)}$ $\left[\frac{\text{N}}{\text{Vm}}\right]$	$\gamma_{2d}^{(1)}$ $\left[\frac{\text{Ns}}{\text{Vm}}\right]$
5	1.136×10^{-11}	-1.11×10^{-10}	285.91	-6.10×10^{16}	7.40×10^{9}	0	3.0×10^{7}	0
10	1.138×10^{-11}	-1.11×10^{-10}	285.91	-5.40×10^{16}	8.50×10^{9}	0	3.0×10^{7}	0
15	1.136×10^{-11}	-1.11×10^{-10}	285.91	-5.85×10^{16}	9.05×10^{9}	0	3.0×10^{7}	0
20	1.136×10^{-11}	-1.11×10^{-10}	285.91	-5.71×10^{16}	9.05×10^{9}	0	3.0×10^{7}	0

Table 5.4: Identified parameters for PIC 181.

A similar optimization of the parameters was done in [117] using both displacement and electric current amplitude responses. For the material PIC 255, applying nonlinear parameters determined from the responses at 30 V to the case of 20 V resulted in a small difference between modeling and experiment in both resonance frequency and amplitude. In other words, a dependence of the elastic modulus and the damping upon the excitation amplitude were exhibited. This can here be verified when a set of parameters is determined for each excitation voltage by fitting the corresponding experimental displacement amplitude response. The piezoelectric parameters γ_0 and $\gamma_2^{(1)}$ are held to be constant.

Tables 5.3 and 5.4 contain the identified parameters of PIC 255 and PIC 181 respectively. The determined values for s_{11}^E and d_{31} also approximate to those of the manufacturer. In table 5.3, the linear elastic parameter s_{11}^E of PIC 255 increases with increasing the excitation voltage, i.e. the linear elastic modulus $E^{(0)}$ decreases. This may be a characteristic of piezoceramic materials, but a heating of piezoceramics after consecutive dynamic experiments close to resonance can also account for the increase in their compliance. In addition, the cubic parameters $E^{(2)}$ and $E_d^{(2)}$ also decrease in absolute value. The changes in parameters of PIC 181 is unclear due to the imperfect shape of resonance curves shown in figures 5.16–5.18.

A very good coincidence between theoretical and experimental results is illustrated in figures 5.11–5.18. It can be seen that the material PIC 255 possesses much higher damping than PIC 181, so that the typical jump phenomenon of Duffing oscillator is suppressed.

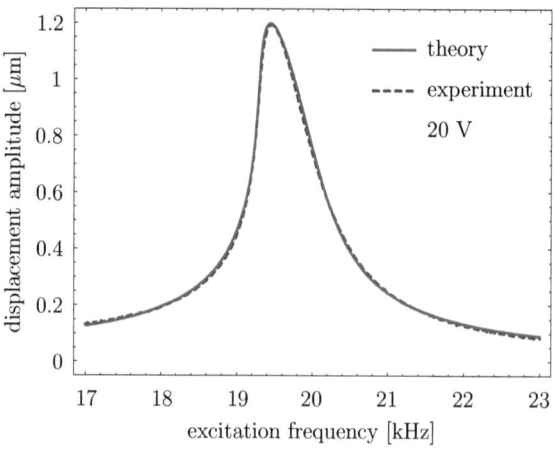

Figure 5.11: Nonlinear parameters fit for PIC 255 at 20 V.

Chapter 5. Nonlinear dynamic modeling 43

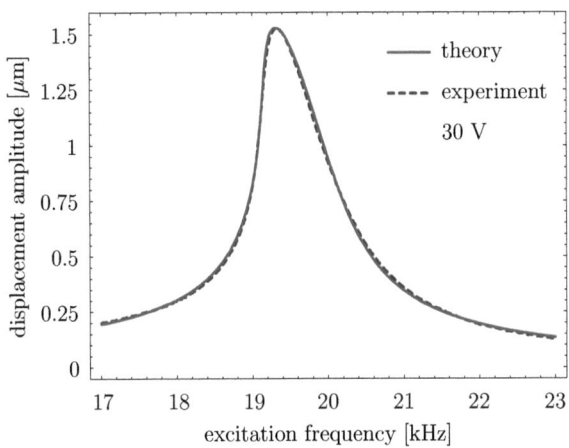

Figure 5.12: Nonlinear parameters fit for PIC 255 at 30 V.

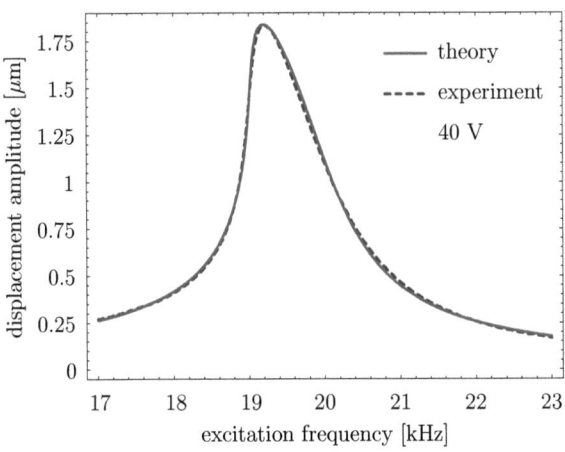

Figure 5.13: Nonlinear parameters fit for PIC 255 at 40 V.

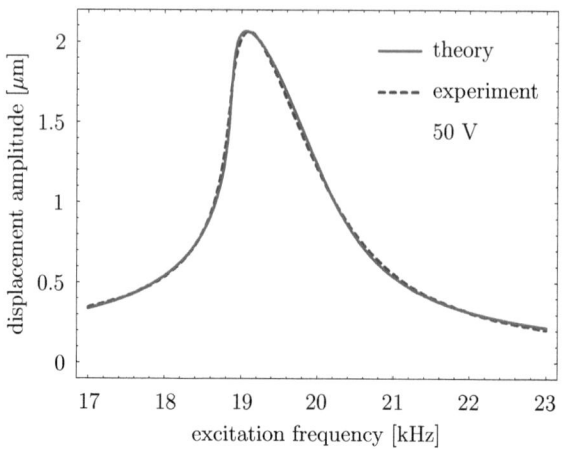

Figure 5.14: Nonlinear parameters fit for PIC 255 at 50 V.

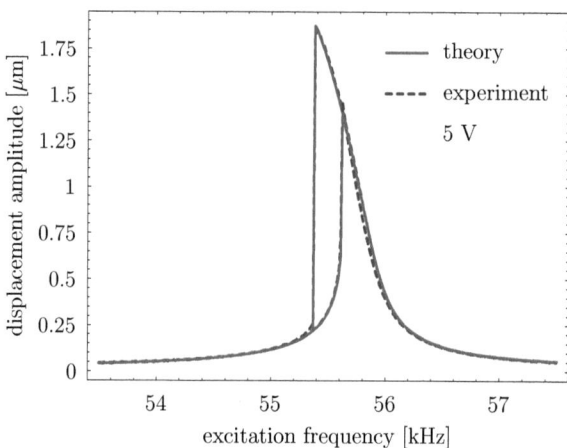

Figure 5.15: Nonlinear parameters fit for PIC 181 at 5 V.

Chapter 5. Nonlinear dynamic modeling 45

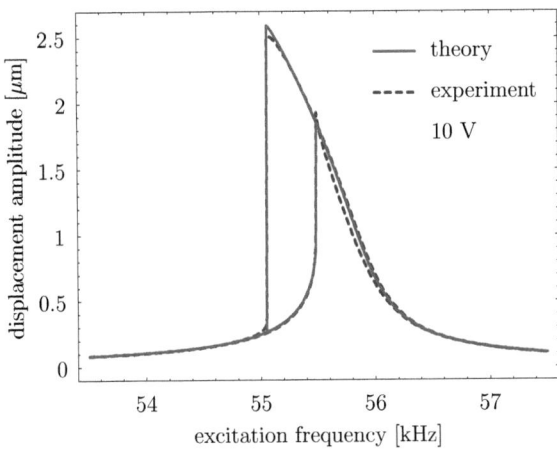

Figure 5.16: Nonlinear parameters fit for PIC 181 at 10 V.

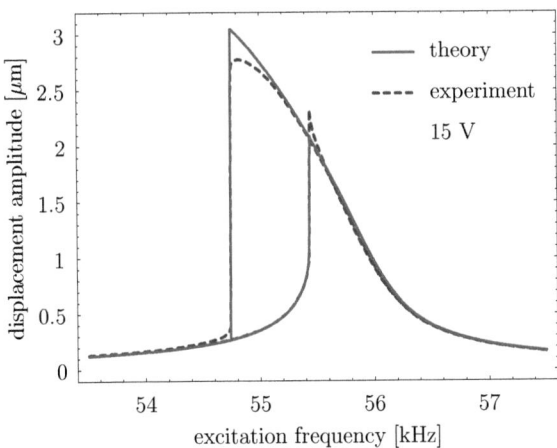

Figure 5.17: Nonlinear parameters fit for PIC 181 at 15 V.

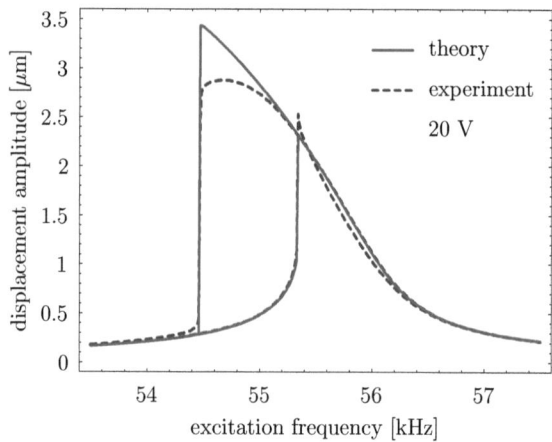

Figure 5.18: Nonlinear parameters fit for PIC 181 at 20 V.

5.5 Consideration for quadratic nonlinearities

Based on the experimental results, that the superharmonic of comparable order of magnitude at the first resonance frequency can be observed when the piezoceramics are excited at one half and one third of the first eigenfrequency [90,117], both quadratic and cubic nonlinearities in the equation (5.13) were taken of the same order ε. As a result, the quadratic nonlinearities were missing in the solution and they could not be determined.

However, it is indicated in [87] that a quadratic nonlinearity can also result in a softening effect on the frequency response curves if the quadratic nonlinearity is considered of order ε and the damping, the cubic nonlinearity and the excitation are taken of order ε^2. A nondimensional equation of motion similar to (5.13) can be derived as

$$\begin{aligned}
& p'' + \varepsilon^2 \, k \, p' + p + \varepsilon \, \alpha_1^* \, p^2 + \varepsilon^2 \, \alpha_{1d}^* \, p \, p' + \varepsilon^2 \, \alpha_1 \, p^3 + \varepsilon^2 \, \alpha_{1d} \, p^2 \, p' \\
& = \varepsilon^2 \, q \, (h_1 \cos \eta\tau + h_2 \sin \eta\tau) + \varepsilon^2 \, q_d \, \eta(-h_2 \cos \eta\tau + h_1 \sin \eta\tau) \\
& \quad + \varepsilon^2 \, \alpha_2^* \, p \, (h_1 \cos \eta\tau + h_2 \sin \eta\tau) + \varepsilon^2 \, \alpha_{2d}^* \, p' \, (h_1 \cos \eta\tau + h_2 \sin \eta\tau) \\
& \quad - \varepsilon^2 \, \alpha_{2d}^* \, p \, \eta(-h_2 \cos \eta\tau + h_1 \sin \eta\tau) + \varepsilon^3 \, \alpha_3^* \, (h_1 \cos \eta\tau + h_2 \sin \eta\tau)^2 \\
& \quad - \varepsilon^3 \, \alpha_{3d}^* \, \eta \, (h_1 \cos \eta\tau + h_2 \sin \eta\tau)(-h_2 \cos \eta\tau + h_1 \sin \eta\tau) \\
& \quad + \varepsilon^2 \, \alpha_2 \, p^2 \, (h_1 \cos \eta\tau + h_2 \sin \eta\tau) + 2\,\varepsilon^2 \, \alpha_{2d} \, p \, p' \, (h_1 \cos \eta\tau + h_2 \sin \eta\tau) \\
& \quad - \varepsilon^2 \, \alpha_{2d} \, p^2 \, \eta(-h_2 \cos \eta\tau + h_1 \sin \eta\tau) \\
& \quad + \varepsilon^3 \, \alpha_3 \, p \, (h_1 \cos \eta\tau + h_2 \sin \eta\tau)^2 + \varepsilon^3 \, \alpha_{3d} \, p' \, (h_1 \cos \eta\tau + h_2 \sin \eta\tau)^2 \\
& \quad - 2\,\varepsilon^3 \, \alpha_{3d} \, p \, \eta(h_1 \cos \eta\tau + h_2 \sin \eta\tau)(-h_2 \cos \eta\tau + h_1 \sin \eta\tau) \\
& \quad + \varepsilon^4 \, \alpha_4 \, (h_1 \cos \eta\tau + h_2 \sin \eta\tau)^3 \\
& \quad - \varepsilon^4 \, \alpha_{4d} \, \eta(h_1 \cos \eta\tau + h_2 \sin \eta\tau)^2(-h_2 \cos \eta\tau + h_1 \sin \eta\tau),
\end{aligned} \qquad (5.28)$$

Chapter 5. Nonlinear dynamic modeling

where

$$()' = \frac{\mathrm{d}}{\mathrm{d}\tau}, \quad k = \frac{d}{\varepsilon^2 m_k \omega_0}, \quad \alpha_1^* = \frac{c_k^{(2)}}{\varepsilon m_k \omega_0^2}, \quad \alpha_{1d}^* = \frac{c_{kd}^{(2)}}{\varepsilon^2 m_k \omega_0},$$

$$\alpha_1 = \frac{c_k^{(3)}}{\varepsilon^2 m_k \omega_0^2}, \quad \alpha_{1d} = \frac{c_{kd}^{(3)}}{\varepsilon^2 m_k \omega_0}, \quad q = \frac{f_k^{(1)}}{\varepsilon^2 m_k \omega_0^2}, \quad q_d = -\frac{f_{kd}^{(1)}}{\varepsilon^2 m_k \omega_0},$$

$$\alpha_2^* = \frac{f_k^{(2)}}{\varepsilon^2 m_k \omega_0^2}, \quad \alpha_{2d}^* = \frac{f_{kd}^{(2)}}{\varepsilon^2 m_k \omega_0}, \quad \alpha_3^* = \frac{f_k^{(3)}}{\varepsilon^3 m_k \omega_0^2}, \quad \alpha_{3d}^* = \frac{f_{kd}^{(3)}}{\varepsilon^3 m_k \omega_0},$$

$$\alpha_2 = \frac{f_k^{(4)}}{\varepsilon^2 m_k \omega_0^2}, \quad \alpha_{2d} = \frac{f_{kd}^{(4)}}{\varepsilon^2 m_k \omega_0}, \quad \alpha_3 = \frac{f_k^{(5)}}{\varepsilon^3 m_k \omega_0^2}, \quad \alpha_{3d} = \frac{f_{kd}^{(5)}}{\varepsilon^3 m_k \omega_0},$$

$$\alpha_4 = \frac{f_k^{(6)}}{\varepsilon^4 m_k \omega_0^2}, \quad \alpha_{4d} = \frac{f_{kd}^{(6)}}{\varepsilon^4 m_k \omega_0}.$$

Applying the Lindstedt-Poincaré method again with the appoximate solution

$$p(\tau) = p_0(\tau) + \varepsilon\, p_1(\tau) + \varepsilon^2 p_2(\tau) + \cdots \tag{5.29}$$

and another expansion for η^2 as

$$\eta^2 = 1 + \varepsilon^2 \delta + \cdots, \tag{5.30}$$

where the term of order ε is omitted so that the corresponding secular term vanishes and it is also reasonable to assume that the deviation of the frequency ratio η from resonance should be $O(\varepsilon^2)$ when the excitation is $O(\varepsilon^2)$ [87, 117].

Substituting (5.29) and (5.30) into the equation (5.28) and equating the coefficients of ε^0, ε^1 and ε^2 on both sides yields

$$\varepsilon^0: \quad p_0'' + \eta^2 p_0 = 0, \tag{5.31}$$

$$\varepsilon^1: \quad p_1'' + \eta^2 p_1 = -\alpha_1^* p_0^2, \tag{5.32}$$

$$\varepsilon^2: \quad p_2'' + \eta^2 p_2 = -k p_0' + \delta p_0 - 2\alpha_1^* p_0 p_1 - \alpha_{1d}^* p_0 p_0' - \alpha_1 p_0^3$$
$$- \alpha_{1d} p_0^2 p_0' + q(h_1 \cos\eta\tau + h_2 \sin\eta\tau) + q_d\, \eta\, (-h_2 \cos\eta\tau + h_1 \sin\eta\tau)$$
$$+ \alpha_2^* p_0 (h_1 \cos\eta\tau + h_2 \sin\eta\tau) + \alpha_{2d}^* p_0' (h_1 \cos\eta\tau + h_2 \sin\eta\tau)$$
$$- \alpha_{2d}^* p_0\, \eta\, (-h_2 \cos\eta\tau + h_1 \sin\eta\tau) + \alpha_2 p_0^2 (h_1 \cos\eta\tau + h_2 \sin\eta\tau)$$
$$+ 2\alpha_{2d} p_0 p_0' (h_1 \cos\eta\tau + h_2 \sin\eta\tau) - \alpha_{2d} p_0^2 \eta\, (-h_2 \cos\eta\tau + h_1 \sin\eta\tau). \tag{5.33}$$

The general solution of the equation (5.31) is the same as that of (5.17) with the unknown amplitude A

$$p_0(\tau) = A \cos\eta\tau. \tag{5.34}$$

Substituting p_0 into (5.32), a stationary solution for p_1 can be derived as

$$p_1(\tau) = -\frac{1}{2\eta^2} \alpha_1^* A^2 \left(1 - \frac{1}{3} \cos 2\eta\tau\right). \tag{5.35}$$

Substituting p_0 and p_1 into (5.33) and eliminating the secular terms yields

$$\delta A - \frac{3}{4}\alpha_1 A^3 + \frac{5}{6\eta^2}\alpha_1^{*2} A^3 + \left(q + \frac{3}{4}\alpha_2 A^2\right) h_1 + \left(-q_d + \frac{1}{4}\alpha_{2d} A^2\right) \eta h_2 = 0, \tag{5.36}$$

$$k\eta A + \frac{1}{4}\alpha_{1d}\eta A^3 + \left(q_d - \frac{3}{4}\alpha_{2d} A^2\right) \eta h_1 + \left(q + \frac{1}{4}\alpha_2 A^2\right) h_2 = 0. \tag{5.37}$$

Taking account of $h_1^2 + h_2^2 = 1$, a fifth order polynomial equation with respect to A^2 can be obtained. It is obvious that the quadratic nonlinearities now have an effect on the zeroth-order response. If the cubic conservative nonlinearity is neglected ($\alpha_1 = 0$), then the corresponding quadratic one always has a softening effect. Otherwise, the effects of the nonlinearities depend on which effect is predominant.

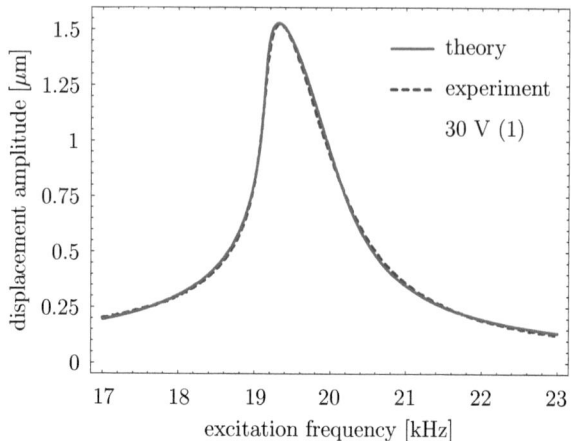

Figure 5.19: Nonlinear parameters fit for PIC 255 at 30 V: Case 1.

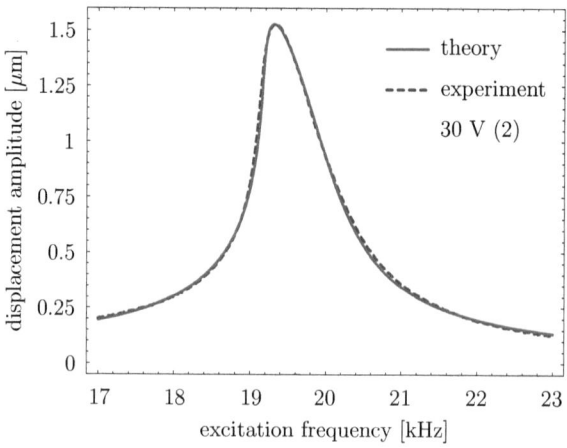

Figure 5.20: Nonlinear parameters fit for PIC 255 at 30 V: Case 2.

Chapter 5. Nonlinear dynamic modeling

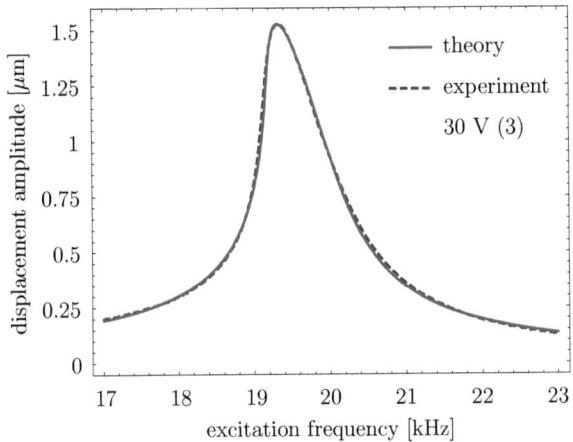

Figure 5.21: Nonlinear parameters fit for PIC 255 at 30 V: Case 3.

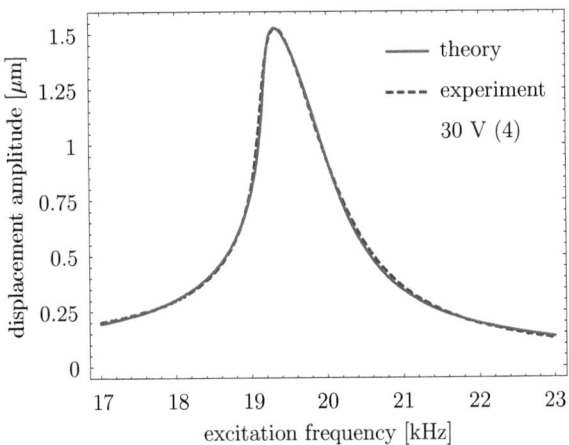

Figure 5.22: Nonlinear parameters fit for PIC 255 at 30 V: Case 4.

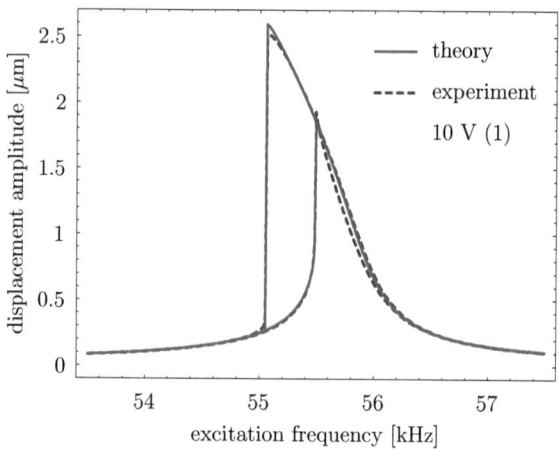

Figure 5.23: Nonlinear parameters fit for PIC 181 at 10 V: Case 1.

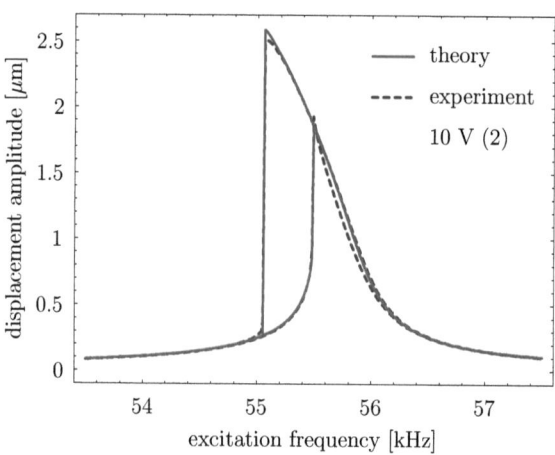

Figure 5.24: Nonlinear parameters fit for PIC 181 at 10 V: Case 2.

Chapter 5. Nonlinear dynamic modeling

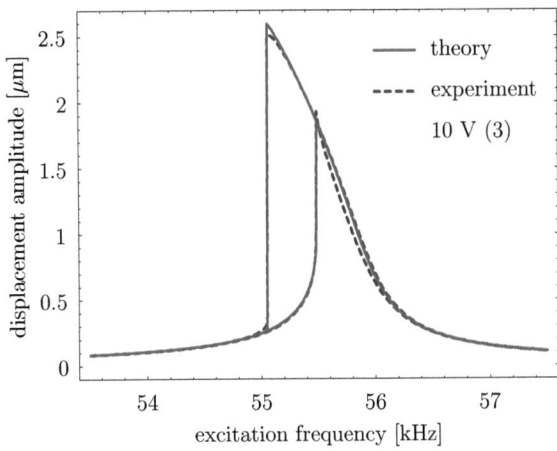

Figure 5.25: Nonlinear parameters fit for PIC 181 at 10 V: Case 3.

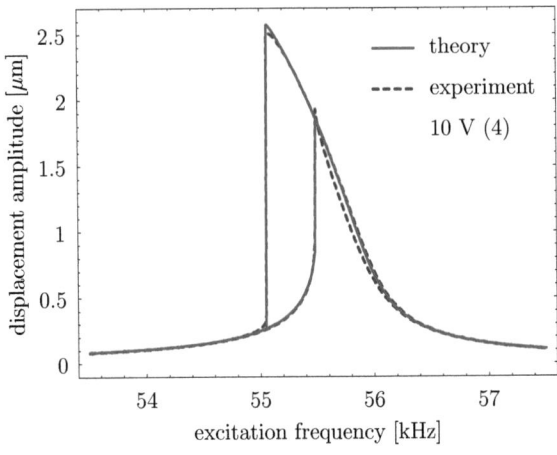

Figure 5.26: Nonlinear parameters fit for PIC 181 at 10 V: Case 4.

Case	$s_{11}^E \left[\frac{m^2}{N}\right]$	$d_{31} \left[\frac{m}{V}\right]$	$E_d^{(0)} \left[\frac{Ns}{m^2}\right]$	$E^{(1)} \left[\frac{N}{m^2}\right]$	$E^{(2)} \left[\frac{N}{m^2}\right]$	$E_d^{(2)} \left[\frac{Ns}{m^2}\right]$	$\gamma_2^{(1)} \left[\frac{N}{Vm}\right]$
1	1.69×10^{-11}	-1.91×10^{-10}	6035	0	-8.02×10^{17}	3.58×10^{12}	1.87×10^{8}
2	1.69×10^{-11}	-1.91×10^{-10}	6035	2.07×10^{14}	0	3.31×10^{12}	9.12×10^{8}
3	1.69×10^{-11}	-1.91×10^{-10}	6035	1.77×10^{14}	-2.12×10^{17}	3.35×10^{12}	8.16×10^{8}
4	1.69×10^{-11}	-1.91×10^{-10}	6035	2.41×10^{14}	2.83×10^{17}	3.31×10^{12}	9.12×10^{8}

Table 5.5: Different sets of identified parameters of PIC 255 at $U_0 = 30$ V with $\gamma_{0d}^{(1)} = \gamma_{2d}^{(1)} = 0$.

Case	$s_{11}^E \left[\frac{m^2}{N}\right]$	$d_{31} \left[\frac{m}{V}\right]$	$E_d^{(0)} \left[\frac{Ns}{m^2}\right]$	$E^{(1)} \left[\frac{N}{m^2}\right]$	$E^{(2)} \left[\frac{N}{m^2}\right]$	$E_d^{(2)} \left[\frac{Ns}{m^2}\right]$	$\gamma_2^{(1)} \left[\frac{N}{Vm}\right]$
1	1.138×10^{-11}	-1.11×10^{-10}	285.91	0	-5.40×10^{16}	8.50×10^{9}	3.00×10^{7}
2	1.138×10^{-11}	-1.11×10^{-10}	285.91	6.59×10^{13}	0	8.50×10^{9}	3.43×10^{7}
3	1.138×10^{-11}	-1.11×10^{-10}	285.91	4.30×10^{13}	-3.10×10^{16}	8.50×10^{9}	2.70×10^{7}
4	1.138×10^{-11}	-1.11×10^{-10}	285.91	7.94×10^{13}	2.41×10^{16}	8.50×10^{9}	4.10×10^{7}

Table 5.6: Different sets of identified parameters of PIC 181 at $U_0 = 10$ V with $\gamma_{0d}^{(1)} = \gamma_{2d}^{(1)} = 0$.

Chapter 5. Nonlinear dynamic modeling

Tables 5.5 and 5.6 contain the different sets of the parameters determined by fitting the experimental displacement amplitude responses for the material PIC 255 and PIC 181, respectively, in four typical cases: (1) – without the quadratic nondissipative mechanical terms, (2) – without the cubic nondissipative mechanical terms, (3) – with both quadratic and cubic nonlinearites having a softening effect, and (4) – with the cubic nonlinearity having a stiffening effect, whereas the quadratic nonlinearity has a softening effect. Figures 5.19–5.26 show a very good coincidence between the theo-retical and experimental results. It can be pointed out that there exist even more possibilities for the identified parameters, leading also to such good coincidence. In other words, the parameter identification is ambiguous, nonlinear behavior of piezoceramics can be accounted for either by the quadratic nonlinearities or by the cubic ones as well as either by mechanical terms or by piezoelectric coupling terms. In the large choice of the parameters, the cubic piezoelectric parameter $\gamma_2^{(1)}$ may be unnecessary at all, as suggested in [117] that the nonlinear elastic terms play a dominant role with the nonlinear behavior of the transversally polarized longitudinal oscillators. To verify this and improve the modeling of the nonlinear dynamic effects, quasi-static behavior of piezoceramics will be considered by performing corresponding experiments in chapter 6.

Chapter 6

Quasi-static experiments

In order to specify the role of the introduced nonlinear or nonconservative terms in the explanation of the nonlinear effects described in chapter 5 and find a more consistent modeling for the dynamic behavior, quasi-static experiments with longitudinal deformations of transversally polarized piezoceramics are performed. The relations between the applied electric field or mechanical stress and the strain as well as the electric displacement density are considered. In contrast to the dynamic experiments described in chapter 3, the electrical and mechanical loads of moderate magnitude are quasi-statically applied to the piezoceramics respectively, so that the corresponding strains are of the same order as those in the dynamic case. Here the excitations are no longer "weak" but it is assumed that they are not "strong" enough to initiate the polarization switching processes. This means the magnitudes of the excitations are still significantly lower than the coercive ones.

As in the dynamic experiments, transversally polarized piezoceramic samples of the materials PIC 255 and PIC 181 manufactured by PI Ceramic are used again. Piezoceramic rods of PIC 255 will be subjected to moderate quasi-static electric fields, whereas those of PIC 181 are used for tension and compression tests under moderate mechanical stresses.

6.1 Experiments with moderate electric field

First, nonlinear behavior of piezoceramics under moderate quasi-static electric fields is experimentally investigated. Experiments are performed at the Chair of High Voltage Technology, Department of Energy and Automation Technology, Technische Universität Berlin (TUB). A free transversally polarized piezoceramic rod as shown in figure 3.1 is excited to longitudinal vibrations by a quasi-static excitation voltage far from resonance, using the inverse 31-effect. The experiments are carried out at 60 Hz on piezoceramic samples of PIC 255 with the dimensions $70 \times 25 \times 3.3$ mm^3, whose first eigenfrequency is about 20 kHz. The longitudinal strains S_{xx} and the electric displacement density D_z in the direction of the excitation electric fields can be obtained with the help of the experimental setup illustrated in figures 6.1 and 6.2. This setup is quite similar to that of the dynamic experiments described in section 3.1.

56   Chapter 6. Quasi-static experiments

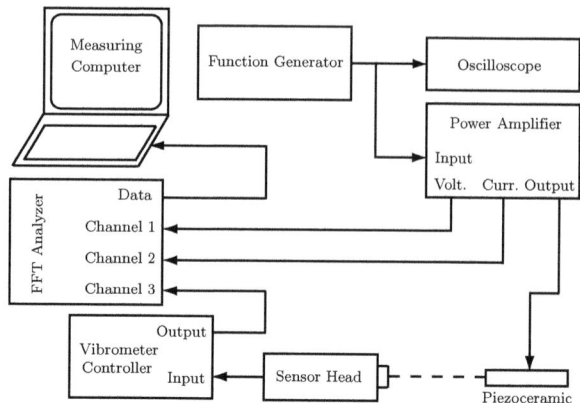

Figure 6.1: Schematic setup of experiments under moderate electric fields.

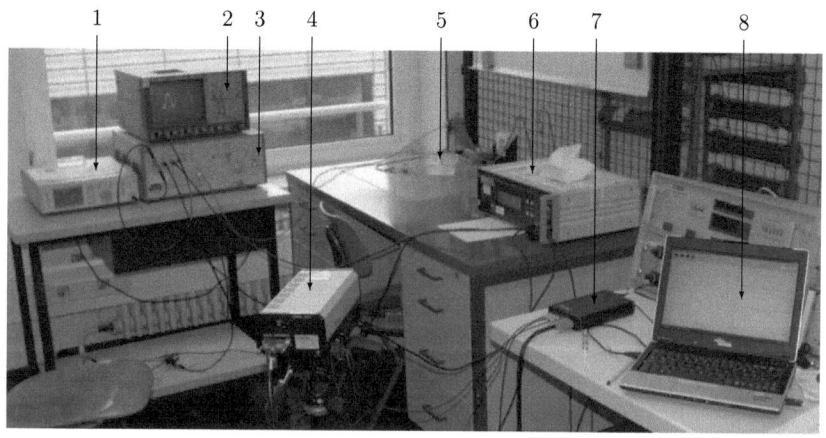

Figure 6.2: Experimental setup of experiments under moderate electric fields: function generator (1), oscilloscope (2), high voltage power amplifier (3), sensor head (4), polarized piezoceramic rod located on an insulated foundation (5), vibrometer controller (6), FFT analyzer (7) and measuring computer (8).

Chapter 6. Quasi-static experiments

The excitation signals are generated by a function generator (Philips PM-5318) and sent to the piezoceramics through a high voltage power amplifier (Trek 10/10A). An oscilloscope (LeCroy 9384AL) is used to monitor the excitation voltages. The mechanical responses are also derived as velocities at one end of the piezoceramics detected by the laser vibrometer. The corresponding displacements can numerically be calculated by integrating the velocity signals with respect to time. Then the longitudinal strains are determined as

$$S_{xx}(t) = \frac{2}{l} u\left(\pm\frac{l}{2}, t\right), \tag{6.1}$$

where l is the length of piezoceramic samples and $u(\pm l/2, t)$ denotes the displacements at the ends of the piezoceramics.

On the electric side, the monitored voltage $U_{mon}(t)$ and the current $I(t)$ through the piezoceramics can directly be obtained as the outputs of the power amplifier. With the assumption of a homogeneous electric field in the z-direction between the electrodes the electric displacement density in this direction is calculated from the equation (5.26) as

$$D_z(t) = \frac{1}{F} \int_0^t I(\tau)\,\mathrm{d}\tau, \tag{6.2}$$

where F is the area of each electrode. All measuring data are finally acquired by a fast Fourier transform (FFT) analyzer and then recorded in a measuring computer.

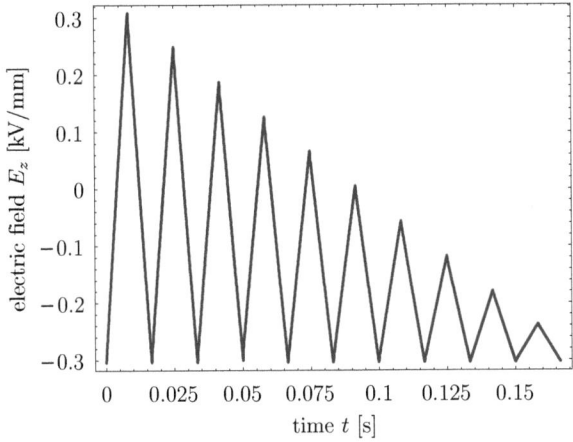

Figure 6.3: Electric field $E_z(t)$ at 60 Hz applied to piezoceramics PIC 255.

In experiments the electric fields in the form of triangular signal as shown in figure 6.3 are used to excite the piezoceramics. Using the samples with a thickness of 3.3 mm, the maximum applied voltage in absolute value is 1000 V corresponding to the electric field of about 0.3 kV/mm which is only 25% of the coercive field for PIC 255 given by the manufacturer $E_c = 1.2$ kV/mm. The results obtained from such experiments for the longitudinal strains and the electric displacements are plotted in figures 6.4 and 6.5

respectively. It is obvious that under moderate electric fields the piezoceramics exhibit a nonlinear hysteretic behavior in both mechanical and electrical responses.

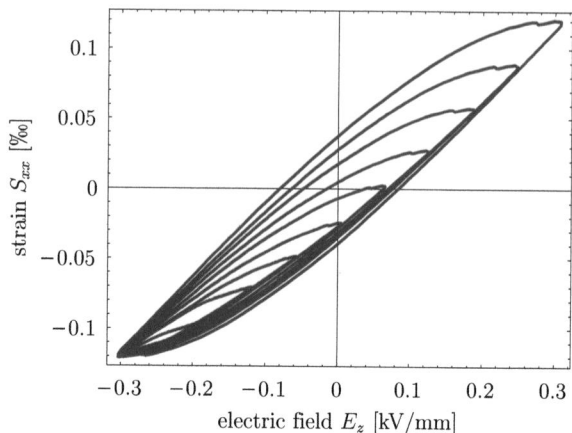

Figure 6.4: Mechanical hysteresis – longitudinal strain S_{xx} vs. applied electric field E_z at 60 Hz for PIC 255.

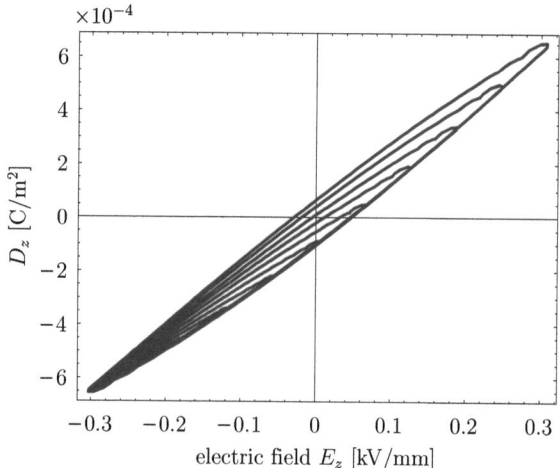

Figure 6.5: Dielectric hysteresis – electric displacement density D_z vs. applied electric field E_z at 60 Hz for PIC 255.

6.2 Experiments with moderate mechanical stress

To investigate the pure mechanical behavior of piezoceramics in the same range of moderate strains occuring when they are subjected to weak electric fields close to resonance, the tension and compression tests are performed on transversally polarized piezoceramic rods of the material PIC 181 with the dimensions 30 × 3 × 2 mm³. Experiments are done at the Chair of Continuum Mechanics and Material Theory, Department of Applied Mechanics, TUB. Figure 6.6 presents a principle of such experiments, where the moderate stresses are quasi-statically applied in the main deformation direction x perpendicular to the polarization direction of the piezoceramics. The longitudinal stresses and strains can be derived by means of the experimental setup shown in figure 6.7.

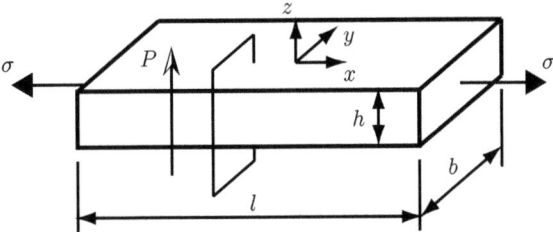

Figure 6.6: Transversally polarized piezoceramic rod subjected to longitudinal stress.

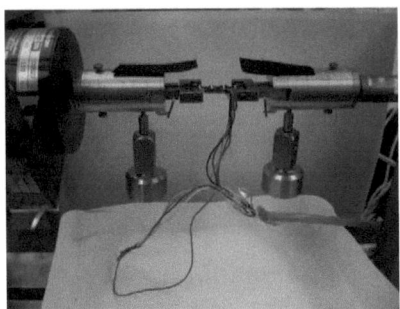 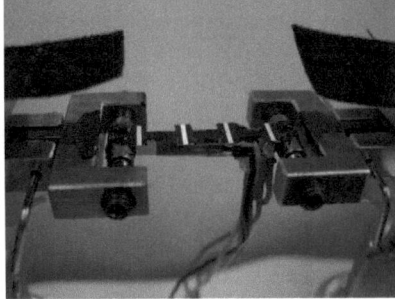

Figure 6.7: Experimental setup of tension and compression tests.

Excluding the influence of coupling and dielectric parameters, the samples are shortcircuited, i.e. no electric field occurs in the piezoceramics. A microforce testing system (Tytron 250) with MTS TestStar IIs controller is used to generate an uniaxial force up to 250 N applied to the piezoceramic rods in the x-direction. This excitation is monitored by a force transducer (MTS 661.11B-02). The strains in the direction of the stresses can be measured either by using a laser extensometer of the type parallel scanner (Fiedler P-50) or by using strain gauges. Due to the necessity to reach a moderate stress with restriction of the excitation force, the cross-section of the samples should be small enough, only 6 mm² in this work. Therefore it is quite difficult to ensure, that a sample is always subjected to a concentric axial load during the tests and thus special clamping devices for both ends of the sample have to be built. As shown in figure 6.7, the piezoceramic rod is fixed at each end by two cylinders connected with two balls. The clamping force is generated by the screws from two sides. Using these clamps, the bending moment acting on the sample (if any) can be minimized. The sample with clamps is then suspended from the testing system so that it can have three rotating degrees of freedom. In addition, for the use of the laser extensometer the upper side of the rod is painted over in black color and four white stripes are sticked on this surface. The strains will be found from a change in the distance of these stripes detected by a laser head.

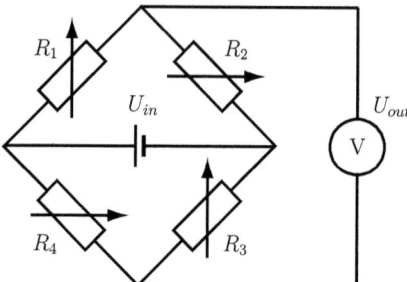

Figure 6.8: Wheatstone full bridge circuit.

It can be pointed out in experiments that the results from the laser extensometer are unstable due to the difficulty in orienting the samples. Hence, the measurements given by strain gauges are only used in this work. The measuring principle using strain gauges is based on the Wheatstone full bridge circuit shown in figure 6.8, which contains four strain gauges as variable resistors, a direct current (DC) voltage source and a voltmeter. Two active strain gauges R_1 and R_3 are glued to a main sample along its length on two opposite sides, while the other strain gauges R_2 and R_4 are the same way glued to another free sample for temperature compensation. Because the four strain gauges (HBM 3/350LY11) have the same gauge factor $k = 2.01$ and the same resistance $R = 350\ \Omega$, the following measurement can be derived [48]

$$\frac{U_{out}}{U_{in}} = \frac{k}{4}\left(S_1 - S_2 + S_3 - S_4\right), \tag{6.3}$$

where U_{out} and U_{in} correspond to the output and input voltage of the bridge and S_i ($i = 1, \ldots, 4$) are the strains of the gauges respectively. In the ideal case, $S_1 = S_3 = S_{xx}$ and $S_2 = S_4 = 0$ the measuring quantity becomes

$$\frac{U_{out}}{U_{in}} = \frac{k}{2} S_{xx}. \tag{6.4}$$

Thus the longitudinal strain can be obtained as

$$S_{xx} = \frac{2}{k}\frac{U_{out}}{U_{in}}. \tag{6.5}$$

Chapter 6. Quasi-static experiments 61

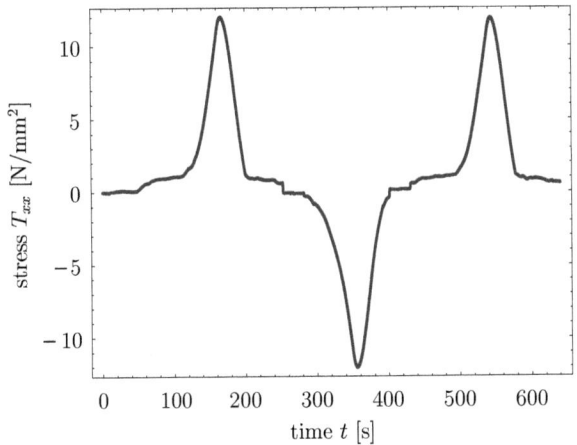

Figure 6.9: Tension and compression stress applied to PIC 181.

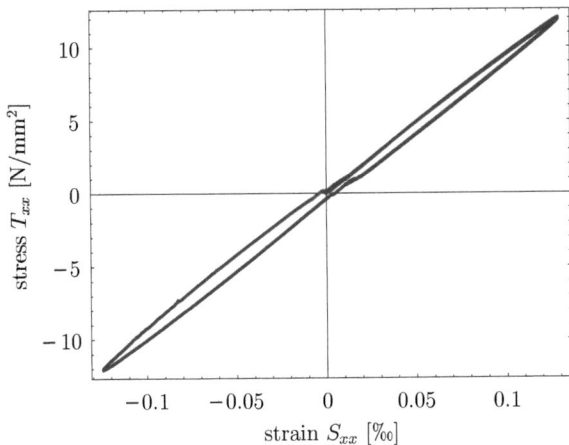

Figure 6.10: Stress–strain hysteresis from tension and compression test for PIC 181.

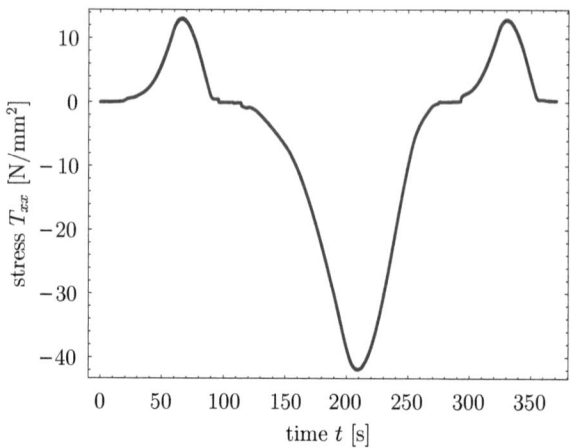

Figure 6.11: Tension and compression stress applied to PIC 181.

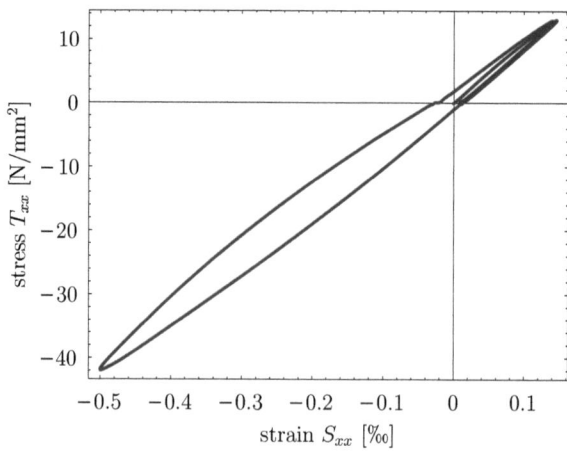

Figure 6.12: Stress–strain hysteresis from tension and compression test for PIC 181.

Chapter 6. Quasi-static experiments

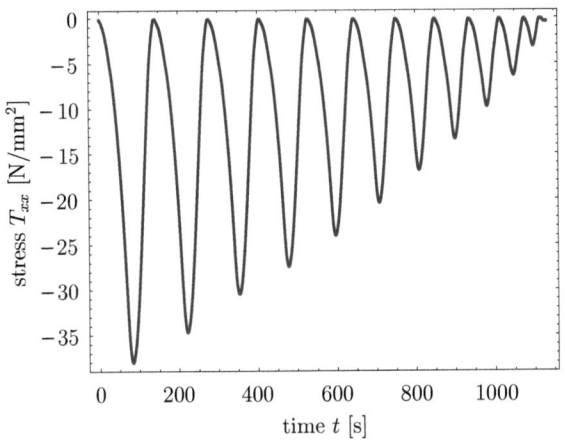

Figure 6.13: Compression stress applied to PIC 181.

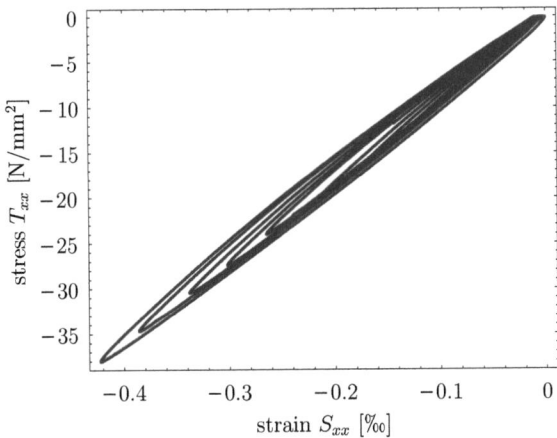

Figure 6.14: Stress–strain hysteresis from compression test for PIC 181.

In experiments, the piezoceramics are subjected to consecutive loading–unloading cycles by tension and/or compression stresses in magnitudes up to about 40 MPa and a maximum loading rate of about 0.5 MPa/s. Figures 6.9, 6.11 and 6.13 show the longitudinal tension and compression stresses applied to the samples over time. The corresponding stress–strain curves are presented in figures 6.10, 6.12 and 6.14 respectively. It is evident that the polarized piezoceramics also exhibit a nonlinear hysteretic behavior under moderate mechanical stresses. In addition, figure 6.10 indicates that the piezoceramics reveal a nearly symmetrical stress–strain relation in this case. Experiments with tension and compression in succession under larger stresses are not successful because it is quite difficult to center precisely the samples on the testing system, so that they often broke at the clamping positions during the tensions. The tension tests always have to be performed very carefully because piezoceramics are brittle materials. Due to the difficulty of the tension tests, only results from the compressions tests will be used in the following.

Chapter 7
Nonlinear quasi-static modeling

This chapter is dedicated to the phenomenological modelings of the nonlinear hysteretic behavior of piezoceramics observed in the quasi-static experiments described in chapter 6. The electrical and mechanical loads are applied to the piezoceramics at low frequencies which are much less than their natural frequency. Therefore this behavior can be considered as a static hysteresis and thus it can be described by rate-independent hysteresis models. In the present work, four of the most common hysteresis models will be used, which are the classical Preisach model, the Prandtl-Ishlinskii model, the Masing model and the Bouc-Wen model respectively.

Experiments show that the hysteretic behavior of the piezoceramics has a nonlocal memory. This means the current value of output response depends not only on the current value of input excitation but also on a history of the input. The classical Preisach model is absolutely appropriate to this type of hysteresis due to its capability to accumulate the past extremum values of the input. In relation with the Preisach model, the Prandtl-Ishlinskii model and its inverse are particular cases having an advantage that the implementation is simpler, because their distribution functions determining the shape of the hysteresis loops can be expressed in analytical forms. The Masing model is structured by a parallel combination of elasto-slide elements as the same as the Prandtl-Ishlinskii model, but the input–output relation is represented by another approach using evolution equations for the internal variables. The last one, Bouc-Wen hysteresis model, can be considered as a generalization of the Masing model. Applying these hysteresis models, a good agreement between theoretical and experimental results can be obtained for both electrical and mechanical cases.

The similar hysteretic behavior of piezoceramics subjected to electric fields has been investigated in several other works. For example, in [44, 45, 126] the classical Preisach model was also used to simulate hysteretic nonlinearities of piezoceramics under strong electric fields (up to 2 kV/mm) initiating polarization switching processes with respect to the inverse 33-effect. The Prandtl-Ishlinskii model can be found in [68, 70] under the name of the Maxwell resistive capacitor model for the description of dielectric hysteresis in piezoceramic transducers with moderate electric fields (less than 80% of the coercive field) using the inverse 31-effect. This model was also applied in [3, 106, 130] describing nonlinear relations between strong excitation electric fields and displacement responses of piezoelectric actuators or nanopositioners. For representing the hysteretic nonlinearity of piezoelectric actuators in control problems, the Bouc-Wen model was utilized e.g. in [38, 75, 123].

On the mechanical side, analogue tension and compression or bending tests at strong stresses in the range from −400 to 80 MPa were considered for instance by [31–34, 131, 132], where "soft" piezoceramics were subjected to loading–unloading cycles. The nonlinear stress–strain behavior can then be explained by domain switching but no hysteresis was taken into account.

7.1 Hysteresis models

7.1.1 Classical Preisach model

The first hysteresis model used in this work was originally proposed for ferromagnetic effects by Preisach [95] and independently developed by Everett and coworkers [27–30] for adsorption of gases by porous solids. An engineering description of this model can be found in a monograph of Mayergoyz [81].

A hysteresis system can be considered as a transducer which is characterized by an input $u(t)$ and a corresponding output $f(t)$. For piezoceramics $u(t)$ is the applied electric field, while $f(t)$ is the electric polarization or $u(t)$ is the stress and $f(t)$ is the strain. The classical Preisach model is based on a simplest hysteresis operator $\widehat{\gamma}_{\alpha\beta}$ represented by a rectangular loop (also called two-position relay) in figure 7.1. Thresholds α and β are "up" and "down" switching values of input, respectively, with the physical assumption that $\alpha \geq \beta$. Output of this elementary hysteresis operator is normally assumed to take only two values +1 and −1 corresponding to "up" and "down" positions, but it may also take two value +1 and 0 [36, 66].

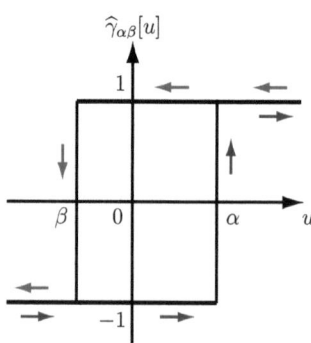

Figure 7.1: Elementary hysteresis operator.

The Preisach model can be defined as a weighted superposition of the elementary operators by [81]

$$f(t) = \widehat{\Gamma}[u(t)] = \iint\limits_{\alpha \geq \beta} \mu(\alpha, \beta)\, \widehat{\gamma}_{\alpha\beta}[u(t)]\, d\alpha\, d\beta, \tag{7.1}$$

where $\mu(\alpha, \beta)$ is the weight function named the Preisach function, which can be identified from experimental data. An interpretation diagram of the Preisach model is shown in figure 7.2 containing infinite elementary operators. But normally, the identification problem can be solved for a discrete approximation to this diagram with finite number of two-position relays. The larger the number of operators used, the better the experimental hysteresis curve is fitted.

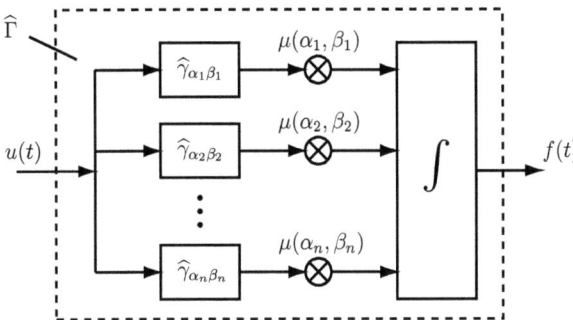

Figure 7.2: Preisach hysteresis model as a continuous weighted superposition of two-position relays ($n \to \infty$).

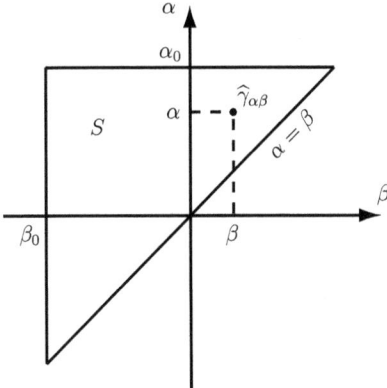

Figure 7.3: The restricted half-plane $\alpha \geq \beta$.

In order to calculate the integral in (7.1), the Preisach model is geometrically interpreted as follows. The two-position operator $\widehat{\gamma}_{\alpha\beta}$ is uniquely characterized by a pair of numbers α and β, which are the switching values of the input $u(t)$, the same as a point of the half-plane $\alpha \geq \beta$ (called the Preisach plane). In practice, the input value is always limited, so the half-plane $\alpha \geq \beta$ can also be restricted to a right triangle S shown in figure 7.3 by a maximum value α_0 of the "up" switching values of the input and a minimum value β_0 of the "down" switching values. This means that the Preisach function $\mu(\alpha, \beta)$ vanishes outside the area S.

Chapter 7. Nonlinear quasi-static modeling

As presented in figure 7.4, starting from the negative boundary state of the hysteresis system when the initial value of the input $u(t_0) \leq \beta_0$, i.e. the input value is less than or equal to the "down" switching values of all $\hat{\gamma}_{\alpha\beta}$-operators in the triangle S, all of them are in the "down" position and all their outputs are equal to -1. The input increases monotonically to some value u_1 at time t_1. At the time $t \in (t_0, t_1)$, $\hat{\gamma}_{\alpha\beta}$-operators, which have the "up" switching value $\alpha < u(t)$, turned into the "up" position and their outputs are equal to $+1$. Thereby, the triangle S is subdivided by the line $\alpha = u(t)$ into two areas $S^+(t)$ and $S^-(t)$ corresponding to the $\hat{\gamma}_{\alpha\beta}$-operators in "up" and "down" position respectively. The dividing line moves upwards until $t = t_1$.

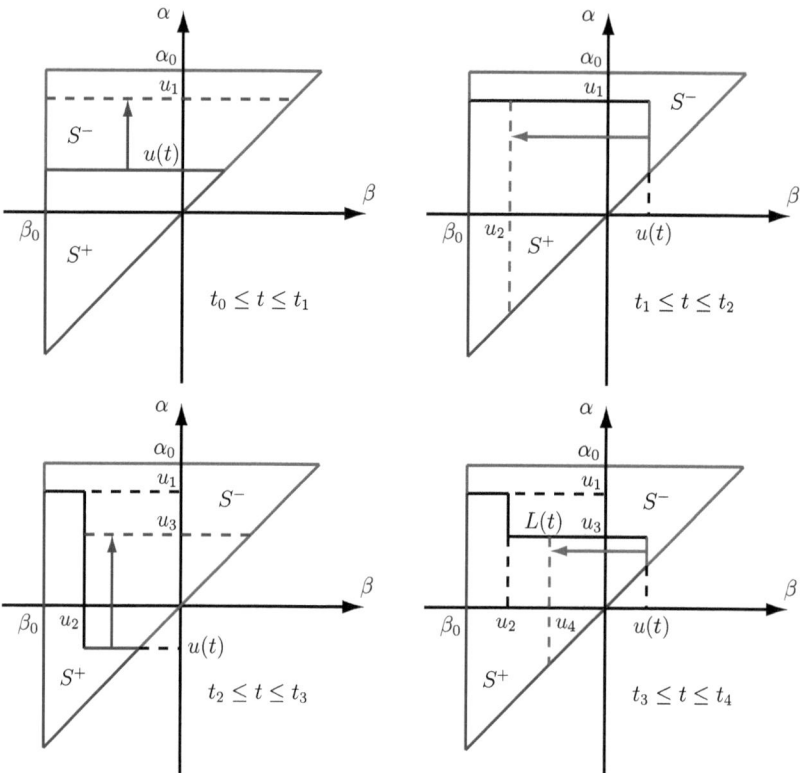

Figure 7.4: The subdivisions of S with respect to the variations of the input: $u(t_0) = \beta_0$ and $u_i = u(t_i)$ with $i = 1, 2, 3, 4$.

Next, the input decreases monotonically to some value u_2 at time t_2. At the time $t \in (t_1, t_2)$, $\hat{\gamma}_{\alpha\beta}$-operators, which have the "down" switching value $\beta > u(t)$, turned into the "down" position and their outputs are equal to -1 again. The triangle S is subdivided into two new areas $S^+(t)$ and $S^-(t)$, between them the interface $L(t)$ now has two segments. The horizontal segment belongs to the line $\alpha = u_1$ and the vertical

Chapter 7. Nonlinear quasi-static modeling   69

one is specified by the line $\beta = u(t)$ which moves to the left until $t = t_2$.

We assume that the input increases monotonically again to some value $u_3 < u_1$ at time t_3 and then decreases monotonically to some value $u_4 > u_2$ at time t_4. Similarly above, this variation of the input will result in a new interface $L(t)$ which now consists of four segments, two horizontal segments are on the lines $\alpha = u_1$ and $\alpha = u_3$, and two vertical others are on the lines $\beta = u_2$ and $\beta = u_4$.

In summary, at any instant of time the limiting triangle S is subdivided into two areas consisting of points (α, β) for which

$$\hat{\gamma}_{\alpha\beta}[u(t)] = \begin{cases} +1 & \text{if } (\alpha, \beta) \in S^+(t), \\ -1 & \text{if } (\alpha, \beta) \in S^-(t). \end{cases} \quad (7.2)$$

Hence, the integral in (7.1) can be subdivided as

$$f(t) = \iint\limits_{S^+(t)} \mu(\alpha, \beta) \, d\alpha \, d\beta - \iint\limits_{S^-(t)} \mu(\alpha, \beta) \, d\alpha \, d\beta. \quad (7.3)$$

Having the above geometric interpretation, it is proven in [81], that a hysteresis system can be represented by the Preisach model if and only if it satisfies both wiping-out and congruency properties. The wiping-out property means that a local maximum of the input wipes out the vertices of the interface $L(t)$ if their α-coordinates are less than this maximum and a local minimum of the input wipes out the vertices whose β-coordinates are greater than this minimum. Thus, only the alternating series of dominant extrema of the input are remaining. For the congruency property, back-and-forth variations of inputs between the same two consecutive extremum values results in hysteresis loops which are geometrically congruent.

It is obvious that the Preisach model is characterized by the weight function $\mu(\alpha, \beta)$ which can be determined from a set of experimental first-order reversal curves. These curves are formed as a monotonic increase of the input from some value less than β_0 is followed by a subsequent monotonic decrease. This variation corresponds to the first reversal of the input. Starting from the negative boundary state of a hysteresis system, the input increases monotonically to some value α and subsequently decreases to some value β as shown in figure 7.5. The corresponding output values are f_α and $f_{\alpha\beta}$ respectively. Defining the Everett function

$$F(\alpha, \beta) = \frac{1}{2}(f_\alpha - f_{\alpha\beta}) = \iint\limits_{T(\alpha,\beta)} \mu(\tilde{\alpha}, \tilde{\beta}) \, d\tilde{\alpha} \, d\tilde{\beta} \quad (7.4)$$

leads to an analytical expression for the Preisach function [81]

$$\mu(\alpha, \beta) = -\frac{\partial^2 F(\alpha, \beta)}{\partial \alpha \, \partial \beta}. \quad (7.5)$$

For the calculation of the output $f(t)$ and the identification of the Preisach function $\mu(\alpha, \beta)$, the corresponding formulas (7.3) and (7.5) should be implemented numerically. Indeed, the zigzag interface $L(t)$ between the positive (S^+) and negative (S^-) areas has vertices whose α- and β-coordinates are elements of the set of the dominant input extrema $\{M_k\}$ and $\{m_k\}$ respectively. The equation (7.3) can be written in the form

$$f(t) = - \iint_S \mu(\alpha,\beta)\,\mathrm{d}\alpha\,\mathrm{d}\beta + 2 \iint_{S^+(t)} \mu(\alpha,\beta)\,\mathrm{d}\alpha\,\mathrm{d}\beta, \qquad (7.6)$$

where $S = S^+(t) \cup S^-(t)$.

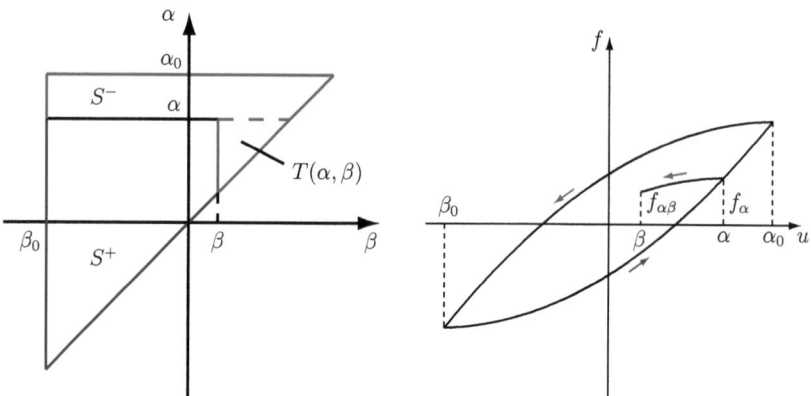

Figure 7.5: The formation of a first-oder reversal curve.

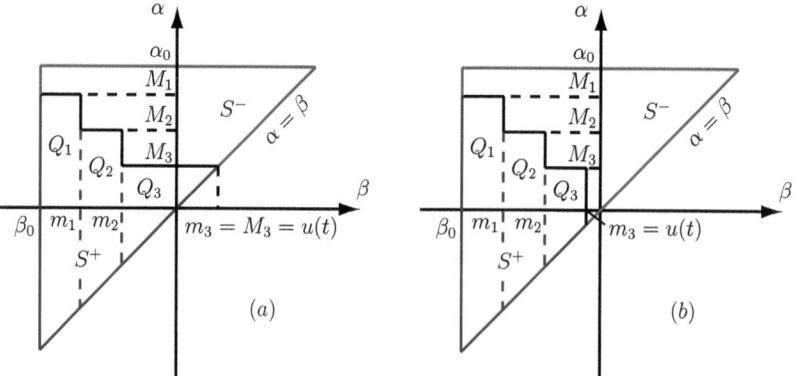

Figure 7.6: Subdivisions of the positive area $S^+(t)$ with $n = 3$.

Subdividing the positive area $S^+(t)$ into n trapezoids Q_k as shown in figure 7.6 and paying attention to the definition (7.4) yields [81]

$$f(t) = -F(\alpha_0,\beta_0) + 2 \sum_{k=1}^{n(t)} \left[F(M_k, m_{k-1}) - F(M_k, m_k) \right], \qquad (7.7)$$

Chapter 7. Nonlinear quasi-static modeling   71

where $m_n = u(t)$ which is the current value of the input and in the case presented in figure 7.6(a) it is valid else that $M_n = u(t)$.

Direct problem

We assume that the input $u(t)$ starts from some value $u(t_0) \leq \beta_0$ and the Preisach function $\mu(\alpha, \beta)$ on the limiting triangle S is given, the output $f(t)$ can easily be evaluated by using the formula (7.7). At any instant of time, a set of sequent dominant extrema of the input $\{M_k, m_k\}$ is determined. Then the values of the Everett function at all vertices of the interface $L(t)$, namely all terms in (7.7), are computed by using the equation (7.4). If only discretized Preisach function is given for a square mesh covering the triangle S as shown in figure 7.7, interpolating function for each cell should be used

$$\mu(\alpha, \beta) = \begin{cases} \mu_0 + \mu_1 \alpha + \mu_2 \beta & \text{for triangle cells,} \\ \mu_0 + \mu_1 \alpha + \mu_2 \beta + \mu_3 \alpha \beta & \text{for square cells,} \end{cases} \quad (7.8)$$

where the coefficients μ_i ($i = 0, 1, 2, 3$) can be found by matching the Preisach function values given at the cell vertices.

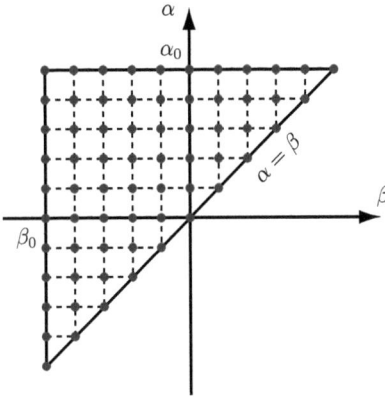

Figure 7.7: A square mesh covering the limiting triangle S.

Instead of the Preisach function, a set of first-order reversal curves can numerically be given. This set comprises the output values $f_{\alpha\beta}$ at the nodes of the mesh. Then the mesh values of the Everett function are also computed by using the formula (7.4) with $f_\alpha \equiv f_{\alpha\alpha}$. The values of the Everett function at the vertices of the interface $L(t)$ can then be interpolated as

$$F(\alpha, \beta) = \begin{cases} F_0 + F_1 \alpha + F_2 \beta & \text{for triangle cells,} \\ F_0 + F_1 \alpha + F_2 \beta + F_3 \alpha \beta & \text{for square cells,} \end{cases} \quad (7.9)$$

where the coefficients F_i ($i = 0, 1, 2, 3$) are determined by matching the mesh values of the Everett function at the cell vertices.

Identification problem

As presented in the direct problem, the Preisach model is specified if a set of first-order reversal curves or the Preisach function is given. Therefore, a solution of the identification problem can be derived directly from experimental results. First, bring the hysteresis system to the negative boundary state by decreasing the input to some value $u(t_0) \leq \beta_0$. Next, a major hysteresis loop can be formed when the input is monotonically increased to the "up" limiting value α_0 and then monotonically decreased to the "down" limiting value β_0 again. The further first-order reversal curves is obtained by the same way, but the input is only increased until it reaches some value α with $\beta_0 < \alpha < \alpha_0$. Of course, all corresponding output values $f_{\alpha\beta}$ are measured.

Another solution of this problem is the Preisach function $\mu(\alpha, \beta)$ which is determined from the above first-order reversal curves. Indeed, by fitting this experimental curves a surface $f_{\alpha\beta}(\alpha, \beta)$ with a given analytic form can be found [39]. Then an analytic expression for the Preisach function is obtained from (7.4) and (7.5) as

$$\mu(\alpha, \beta) = \frac{1}{2} \frac{\partial^2 f_{\alpha\beta}}{\partial \alpha \, \partial \beta}. \qquad (7.10)$$

By a similar method, the Preisach function is assumed as a well-known distribution function, such as factorized Cauchy-Lorentz, lognormal/Gaussian, Gaussian/Gaussian [13, 46], or Cauchy-Lorentz/Gaussian, Cauchy-Lorentz/lognormal [96] distributions. The parameters of the distribution functions are then determined by fitting the experimental major loop. These methods have the advantage that there are only few parameters and the major hysteresis loop is easily measured. However, it can be seen in [46, 96] that even the approximation of the major loop is not so precise and the minor loops are obviously inaccurate.

In another way, the Preisach function can be directly identified with a better precision from experimental data. Two similar algorithms are proposed in [14–16] and [50] respectively. With the first algorithm the weight function can be determined at any point in the Preisach plane for arbitrary accuracy by applying two appropriate input functions around this point. By contrast, the later algorithm calculates simultaneously all values of the weight function in the discretized Preisach plane by applying a finite number of chosen input functions. This number depends on the number of the cells. From the numerical results, the weight function can be approximated as a linear combination of suitable basis functions, where the constant coefficients are computed by a least squares method [49]. These algorithms are very simple to evaluate the discretized Preisach function, but the corresponding measurements are complex and the result is sensitive to experimental errors.

Solving also a constrained least squares problem, the discretized Preisach function can be found by fitting the concatenation of the first-order reversal curves, where the function within each cell in the Preisach plane is assumed to concentrate at the cell center as a discrete mass [54, 107, 110]. Nevertheless, the input has to be discretized so that the vertices of the interface $L(t)$ between positive and negative areas in the Preisach plane always coincide with the nodes of the grid. In other words, it is difficult to use this solution for a general input, of which some dominant extremum value does not correspond to any node of the grid.

Chapter 7. Nonlinear quasi-static modeling

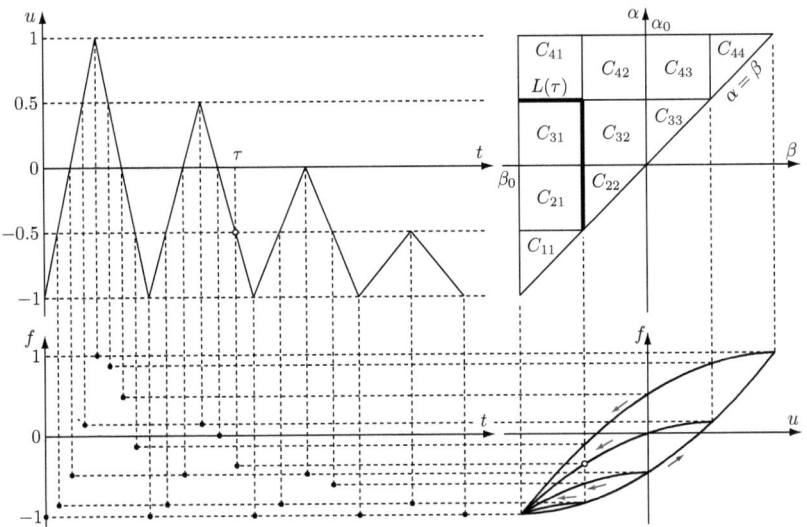

Figure 7.8: Discretization of the Preisach plane for $n = 4$.

In this work, a simple method to identify the Preisach function will be presented as follows. In experiments, a set of n first-order reversal curves is measured. Then the Preisach plane can be discretized into n levels on each axes. Figure 7.8 shows an example for $n = 4$, where the cells are labeled upwards and from left to right. The upper left corner of each cell takes also the index of the cell.

As mentioned above, the mesh values of the Everett function F_{ij} ($i, j = 1, \ldots, n$ and $i \geq j$) can be computed from the first-order reversal curves by using the first equation in (7.4). Then the double integral of the Preisach function over each cell is easily determined

$$I_{ij} = \iint_{C_{ij}} \mu(\alpha, \beta) \, d\alpha \, d\beta. \tag{7.11}$$

The first n integrals over the triangle cells are directly obtained from (7.4)

$$I_{ii} = F_{ii} \quad (i = 1, 2, \ldots, n). \tag{7.12}$$

The next $n - 1$ integrals over the adjacent cells can be derived from the equations (7.4) and (7.11). It is clearly that

$$F_{i+1,i} = I_{i+1,i} + I_{ii} + I_{i+1,i+1} \quad (i = 1, 2, \ldots, n-1). \tag{7.13}$$

Therefore,

$$I_{i+1,i} = F_{i+1,i} - I_{ii} - I_{i+1,i+1} \quad (i = 1, 2, \ldots, n-1). \tag{7.14}$$

The rest cells C_{ij} of the Preisach plane is divided into $n - 2$ levels due to the difference inside their indices, $\Delta i = i - j$. The integrals of the Preisach function over these cells can be found for each level Δi from 2 to $n-1$ in succession. From the equations (7.4) and (7.11),

$$F_{ij} = I_{ij} + F_{i,j+1} + F_{i-1,j} - F_{i-1,j-1}. \tag{7.15}$$

Thus,

$$I_{ij} = F_{ij} - F_{i,j+1} - F_{i-1,j} + F_{i-1,j-1}, \tag{7.16}$$

with $i = \Delta i + 1, \ldots, n$ and $j = i - \Delta i$.

Assume that the discretized Preisach function within each cell in the Preisach plane is constant. From the equation (7.11),

$$I_{ij} = \mu_{ij} A_{ij}, \tag{7.17}$$

where A_{ij} are the areas of the cells and μ_{ij} correspond to $n(n+1)/2$ unknown constants respectively. Hence, these values of the discretized Preisach function can be computed from (7.17) by using the integral values in (7.12), (7.14) and (7.16) as

$$\mu_{ij} = \frac{I_{ij}}{A_{ij}} \quad (i,j = 1, 2, \ldots, n \text{ and } i \geq j). \tag{7.18}$$

Note that the Preisach function should be non-negative, as it plays the role of a density function. This condition is automatically satisfied when the input and the output are in direct variation, that means they increase or decrease together.

More generally the Preisach function in each cells $\mu_{ij}(\alpha, \beta)$ can be assumed to represent a plane or a doubly ruled surface. First, all values of the Preisach function at the nodes on the line $\alpha = \beta$ are determined by the method of Biorci and Pescetti summarized in [50]. For n triangle cells, the Preisach function has the form

$$\mu_{ii}(\alpha, \beta) = \mu_0 + \mu_1 \alpha + \mu_2 \beta \quad (i = 1, 2, \ldots, n), \tag{7.19}$$

where the μ-coefficients are found by fitting the corresponding integral I_{ii} in (7.12) and the two specified vertex values of the Preisach function in the cell.

For the rest cells, the Preisach function has the form

$$\mu_{ij}(\alpha, \beta) = \mu_0 + \mu_1 \alpha + \mu_2 \beta + \mu_3 \alpha \beta, \tag{7.20}$$

with the μ-coefficients are found by fitting the corresponding integral I_{ij} in (7.14) or (7.16) and the three specified vertex values of the Preisach function in the cell.

7.1.2 Prandtl-Ishlinskii model

The second hysteresis model introduced here is a Prandtl-Ishlinskii model of stop-type [116]. It is represented by a parallel combination of elementary stop hysterons, each of them is composed of a linear spring coupled in series with a pure Coulomb friction element. Figure 7.9 shows such a system, where the stiction force H_0 of the zeroth slide element is assumed to be infinitely large. Because this construction was initially proposed by Maxwell, it may be referred to as Maxwell model or Maxwell resistive capacitor model in literature e.g. [37, 68–71].

Chapter 7. Nonlinear quasi-static modeling

The constitutive behavior of the system shown in figure 7.9 can be described by

$$F(t) = c_0 u(t) + \sum_{i=1}^{n} K_i(t), \quad (7.21)$$

$$K_i = \begin{cases} c_i(u - u_i) & \text{if } |K_i| < H_i \text{ or } |K_i| = H_i \wedge \text{sgn}(\dot{u}K_i) \leq 0, \\ H_i \, \text{sgn}(\dot{u}) & \text{otherwise,} \end{cases} \quad (7.22)$$

where u is the input displacement, F is the output force, and K_i, c_i, H_i and u_i are the output force, spring stiffness, stiction force and displacement from initial equilibrium of the i-th elasto-slide element respectively. The friction force is assumed to be equal to the stiction force when a block slips.

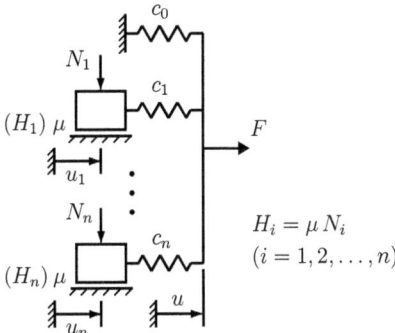

Figure 7.9: Prandtl-Ishlinskii model.

The Prandtl-Ishlinskii (PI) model and its inverse are established to be particular cases of the classical Preisach model described in section 7.1.1 and the Preisach function can be determined in an analytic form [100]. Considering a PI model with $n = 2$ elasto-slide elements and assuming that $H_1 > H_2$, it is easy to construct the relationship of the output $f(t) \equiv F(t)$ to the input $u(t)$ as plotted in figure 7.10. This hysteretic nonlinearity is completely characterized by the parameters c_0, c_1, c_2, w_1 and w_2, with

$$w_i = 2\frac{H_i}{c_i} \quad (i = 1, 2). \quad (7.23)$$

The skeleton lines for the descending and ascending branches are respectively represented by

$$f = c_0 u \pm H_1 \mp H_2 \left(1 + 2\frac{c_1}{c_2}\right). \quad (7.24)$$

In order to find the Preisach function $\mu(\alpha, \beta)$ first-order reversal curves $f_{\alpha\beta}$ should be specified and then the formula (7.10) is applied. It is obvious from figure 7.10 that

$$f_{\alpha\beta} = \begin{cases} (c_0 + c_1 + c_2)\beta + A_1 & \text{if } \alpha - w_2 < \beta \leq \alpha, \\ (c_0 + c_1)\beta + A_2 & \text{if } \alpha - w_1 < \beta \leq \alpha - w_2, \\ c_0 \beta - H_1 - H_2 & \text{if } -\infty < \beta \leq \alpha - w_1, \end{cases} \quad (7.25)$$

where the constants A_1 and A_2 depend on the input values at the first-order reversal points α.

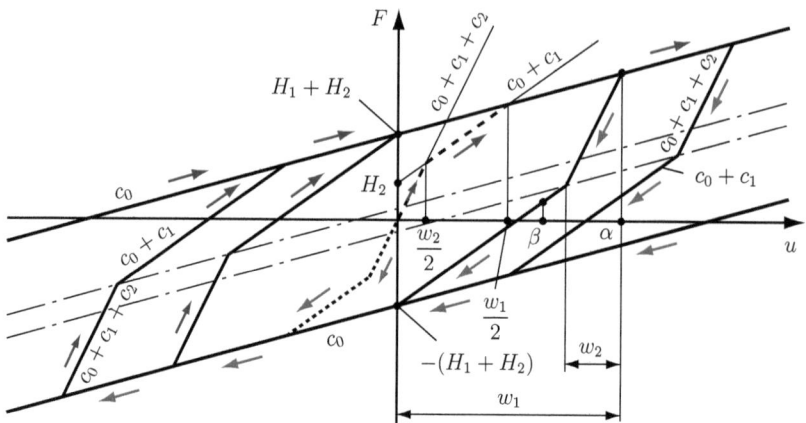

Figure 7.10: Simple Prandtl-Ishlinskii model with $n = 2$.

Hence, the partial derivative of $f_{\alpha\beta}$ with respect to β can be derived as

$$\frac{\partial f_{\alpha\beta}}{\partial \beta} = \begin{cases} c_0 + c_1 + c_2 & \text{if } \beta \leq \alpha < \beta + w_2, \\ c_0 + c_1 & \text{if } \beta + w_2 \leq \alpha < \beta + w_1, \\ c_0 & \text{if } \beta + w_1 \leq \alpha < \infty. \end{cases} \quad (7.26)$$

With the Heaviside step function

$$\theta(x) = \begin{cases} 1 & \text{if } x \geq 0, \\ 0 & \text{otherwise,} \end{cases} \quad (7.27)$$

the above partial derivative can be grouped into the following equation

$$\frac{\partial f_{\alpha\beta}}{\partial \beta} = (c_0 + c_1 + c_2)\,\theta(\alpha - \beta) - c_1\,\theta(\alpha - \beta - w_1) - c_2\,\theta(\alpha - \beta - w_2). \quad (7.28)$$

From the equation (7.10) the Preisach function is obtained as

$$\mu(\alpha, \beta) = \frac{1}{2}\left[(c_0 + c_1 + c_2)\,\delta(\alpha - \beta) - c_1\,\delta(\alpha - \beta - w_1) - c_2\,\delta(\alpha - \beta - w_2)\right], \quad (7.29)$$

where δ denotes the Dirac delta function.

For the case of figure 7.9, the Preisach function can be generalized as [71]

$$\mu(\alpha, \beta) = \frac{1}{2}\left\{\sum_{i=0}^{n} c_i\,\delta(\alpha - \beta) - \sum_{i=1}^{n}\left[c_i\,\delta(\alpha - \beta - w_i)\right]\right\}, \quad (7.30)$$

Chapter 7. Nonlinear quasi-static modeling

with $w_i = 2H_i/c_i$ ($i = 1, 2, \ldots, n$) as given in (7.23). It is obvious that the Preisach function of the PI model consists of the line $\alpha = \beta$ and n lines $\alpha = \beta + w_i$ parallel to this line for a finite number of elasto-slide elements.

For numerical implementation and model identification, the Everett function should be found by using its definition (7.4)

$$F(\alpha,\beta) = \iint_{T(\alpha,\beta)} \mu(\widetilde{\alpha},\widetilde{\beta})\,\mathrm{d}\widetilde{\alpha}\,\mathrm{d}\widetilde{\beta} = \int_\beta^\alpha \left[\int_{\widetilde{\beta}}^\alpha \mu(\widetilde{\alpha},\widetilde{\beta})\mathrm{d}\widetilde{\alpha}\right]\mathrm{d}\widetilde{\beta}. \tag{7.31}$$

Introducing (7.30) into (7.31) yields

$$F(\alpha,\beta) = \frac{1}{2}\left\{\sum_{i=0}^n c_i(\alpha - \beta) - \sum_{i=1}^n [c_i(\alpha - \beta - w_i)\theta(\alpha - \beta - w_i)]\right\}, \tag{7.32}$$

where θ denotes the Heaviside step function. It can be seen that the value of the Everett function of the PI model is constant along lines parallel to the line $\alpha = \beta$. Unlike the Preisach function, $F(\alpha,\beta)$ is a continuous function over the limiting triangle S in the Preisach plane.

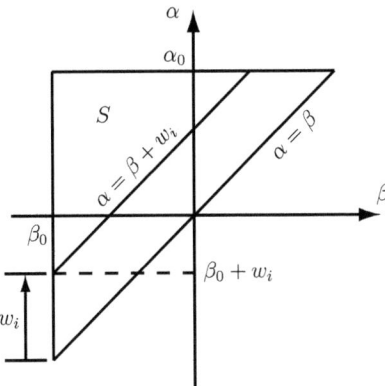

Figure 7.11: Prandtl-Ishlinskii model identification.

Since the PI model is a particular case of the classical Preisach model, it can be identified by using the Everett function values calculated from the experimental first-order reversal curves. The limiting triangle S is now divided by $n+1$ lines $\alpha = \beta + w_i$ ($i = 0, 1, \ldots, n$) as shown in figure 7.11. It is possible to average the values of the experimental Everett function along each of these lines. These averaged values can be expressed as

$$F_{\alpha i} = \frac{1}{\alpha_0 - \beta_0 - w_i}\int_{\beta_0 + w_i}^{\alpha_0} F(\alpha, \alpha - w_i)\,\mathrm{d}\alpha. \tag{7.33}$$

From the equations (7.32) and (7.33),

$$F_{\alpha i} = \frac{1}{2}\left(\sum_{k=0}^i w_i c_k + \sum_{k=i+1}^n w_k c_k\right) \qquad (i = 0, 1, \ldots, n). \tag{7.34}$$

Giving the distances w_i and the corresponding values $F_{\alpha i}$ the unknown stiffnesses c_i can be solved by

$$\mathbf{c} = 2 \begin{bmatrix} w_0 & w_1 & w_2 & \cdots & w_{n-1} & w_n \\ w_1 & w_1 & w_2 & \cdots & w_{n-1} & w_n \\ w_2 & w_2 & w_2 & \cdots & w_{n-1} & w_n \\ \vdots & \vdots & \vdots & \ddots & \vdots & \vdots \\ w_{n-1} & w_{n-1} & w_{n-1} & \cdots & w_{n-1} & w_n \\ w_n & w_n & w_n & \cdots & w_n & w_n \end{bmatrix}^{-1} \mathbf{F}_\alpha, \qquad (7.35)$$

where $\mathbf{c} = [c_0, c_1, \ldots, c_n]^T$ and $\mathbf{F}_\alpha = [F_{\alpha 0}, F_{\alpha 1}, \ldots, F_{\alpha n}]^T$. Then the corresponding stiction forces can be determined from (7.23)

$$H_i = \frac{1}{2} c_i w_i \qquad (i = 1, 2, \ldots, n). \qquad (7.36)$$

In a special case of the symmetric limiting triangle S with $|\alpha_0| = |\beta_0|$ and the values w_i decrease equally in length

$$w_i = 2\alpha_0 \left(1 - \frac{i}{n+1}\right) \qquad (i = 0, 1, \ldots, n), \qquad (7.37)$$

the equations (7.35) and (7.36) become

$$\mathbf{c} = \frac{n+1}{\alpha_0} \begin{bmatrix} n+1 & n & n-1 & \cdots & 2 & 1 \\ n & n & n-1 & \cdots & 2 & 1 \\ n-1 & n-1 & n-1 & \cdots & 2 & 1 \\ \vdots & \vdots & \vdots & \ddots & \vdots & \vdots \\ 2 & 2 & 2 & \cdots & 2 & 1 \\ 1 & 1 & 1 & \cdots & 1 & 1 \end{bmatrix}^{-1} \mathbf{F}_\alpha$$

$$= \frac{n+1}{\alpha_0} \begin{bmatrix} 1 & -1 & 0 & \cdots & & 0 \\ -1 & 2 & -1 & \ddots & \ddots & \vdots \\ 0 & -1 & 2 & -1 & \ddots & \vdots \\ \vdots & \ddots & -1 & 2 & -1 & 0 \\ \vdots & \ddots & \ddots & -1 & 2 & -1 \\ 0 & \cdots & \cdots & 0 & -1 & 2 \end{bmatrix} \mathbf{F}_\alpha \qquad (7.38)$$

and

$$H_i = c_i \alpha_0 \left(1 - \frac{i}{n+1}\right) \qquad (i = 1, 2, \ldots, n). \qquad (7.39)$$

The inverse Prandtl-Ishlinskii model

We consider now the PI model given in figure 7.9 but with the input $F(t)$ and the output $u(t)$. From figure 7.10 permuting the role of input and output, and generalizing for the case of figure 7.9 will lead to the Preisach function of the inverse PI model

$$\overline{\mu}(\overline{\alpha}, \overline{\beta}) = \frac{1}{2} \left\{ \sum_{i=0}^{n} \overline{c}_i \delta(\overline{\alpha} - \overline{\beta}) + \sum_{i=1}^{n} \left[\overline{c}_i \delta(\overline{\alpha} - \overline{\beta} - \overline{w}_i)\right] \right\}, \qquad (7.40)$$

Chapter 7. Nonlinear quasi-static modeling 79

where an overline denotes that it is for the inverse and δ refers to the Dirac delta function,

$$\overline{c}_0 = \frac{2}{\sum_{k=0}^{n} c_k} - \frac{1}{c_0}, \qquad (7.41)$$

$$\overline{c}_i = \frac{1}{\sum_{k=0}^{i-1} c_k} - \frac{1}{\sum_{k=0}^{i} c_k} \quad (i = 1, 2, \ldots, n) \qquad (7.42)$$

and

$$\overline{w}_n = w_n \sum_{k=0}^{n} c_k, \qquad (7.43)$$

$$\overline{w}_{i-1} = \overline{w}_i + (w_{i-1} - w_i) \sum_{k=0}^{i} c_k \quad (i = n, \ldots, 3, 2). \qquad (7.44)$$

Substituting the equation (7.40) into (7.31), the Everett function of the inverse PI model can be derived as

$$\overline{F}(\overline{\alpha}, \overline{\beta}) = \frac{1}{2} \left\{ \sum_{i=0}^{n} \overline{c}_i (\overline{\alpha} - \overline{\beta}) + \sum_{i=1}^{n} \left[\overline{c}_i (\overline{\alpha} - \overline{\beta} - \overline{w}_i) \theta(\overline{\alpha} - \overline{\beta} - \overline{w}_i) \right] \right\}, \qquad (7.45)$$

where θ denotes the Heaviside step function (7.27).

The difference between the inverse and direct PI model is that the Preisach function $\overline{\mu}(\overline{\alpha}, \overline{\beta})$ is non-negative having physically meaning of a density function, whereas $\mu(\alpha, \beta)$ is positive only on the line $\alpha = \beta$. Additionally, in the direction of $(\overline{\alpha}, \overline{\beta}) = (1, -1)$ the slope of the Everett function $\overline{F}(\overline{\alpha}, \overline{\beta})$ increases in value, but that of $F(\alpha, \beta)$ decreases in value in the same direction.

Similar to the direct PI model, the inverse model can be determined from the experimental first-order reversal curves as following

$$\overline{\mathbf{c}} = 2 \begin{bmatrix} \overline{w}_0 & 2\overline{w}_0 - \overline{w}_1 & 2\overline{w}_0 - \overline{w}_2 & \cdots & 2\overline{w}_0 - \overline{w}_{n-1} & 2\overline{w}_0 - \overline{w}_n \\ \overline{w}_1 & \overline{w}_1 & 2\overline{w}_1 - \overline{w}_2 & \cdots & 2\overline{w}_1 - \overline{w}_{n-1} & 2\overline{w}_1 - \overline{w}_n \\ \overline{w}_2 & \overline{w}_2 & \overline{w}_2 & \cdots & 2\overline{w}_2 - \overline{w}_{n-1} & 2\overline{w}_2 - \overline{w}_n \\ \vdots & \vdots & \vdots & \ddots & \vdots & \vdots \\ \overline{w}_{n-1} & \overline{w}_{n-1} & \overline{w}_{n-1} & \cdots & \overline{w}_{n-1} & 2\overline{w}_{n-1} - \overline{w}_n \\ \overline{w}_n & \overline{w}_n & \overline{w}_n & \cdots & \overline{w}_n & \overline{w}_n \end{bmatrix}^{-1} \mathbf{F}_{\overline{\alpha}}, \qquad (7.46)$$

where $0 < \overline{w}_i \leq \overline{\alpha}_0 - \overline{\beta}_0$ are the given distances from the line $\overline{\alpha} = \overline{\beta}$ to lines parallel to that line, $\overline{\mathbf{c}} = [\overline{c}_0, \overline{c}_1, \ldots, \overline{c}_n]^{\mathrm{T}}$ and $\mathbf{F}_{\overline{\alpha}} = [\overline{F}_{\overline{\alpha}0}, \overline{F}_{\overline{\alpha}1}, \ldots, \overline{F}_{\overline{\alpha}n}]^{\mathrm{T}}$, with $\overline{F}_{\overline{\alpha}i}$ are the average values of the experimentally determined Everett function along each of the lines $\overline{\alpha} = \overline{\beta} + \overline{w}_i$,

$$\overline{F}_{\overline{\alpha}i} = \frac{1}{\overline{\alpha}_0 - \overline{\beta}_0 - \overline{w}_i} \int_{\overline{\beta}_0 + \overline{w}_i}^{\overline{\alpha}_0} \overline{F}(\overline{\alpha}, \overline{\alpha} - \overline{w}_i) \, \mathrm{d}\overline{\alpha} \quad (i = 0, 1, \ldots, n). \qquad (7.47)$$

Then from the equations (7.41)–(7.44) the parameters of the direct PI model can be extracted as follows

$$c_0 = \left(2\sum_{k=0}^{n}\overline{c}_k - \overline{c}_0\right)^{-1}, \tag{7.48}$$

$$c_i = \left(1 \Big/ \sum_{k=0}^{i-1}\overline{c}_k - \overline{c}_i\right)^{-1} - \sum_{k=0}^{i-1} c_k \quad (i = 1, 2, \ldots, n), \tag{7.49}$$

$$w_n = \overline{w}_n \Big/ \sum_{k=0}^{n} c_k, \tag{7.50}$$

$$w_{i-1} = w_i + (\overline{w}_{i-1} - \overline{w}_i) \Big/ \sum_{k=0}^{i} c_k \quad (i = n, \ldots, 3, 2) \tag{7.51}$$

and the stiction forces $H_i = \frac{1}{2} c_i w_i$ $(i = 1, 2, \ldots, n)$ as given in (7.36).

7.1.3 Masing model

As the third applied hysteresis model, the Masing model proposed in [78] will be here presented. The structure of this model is the same as that of the Prandtl-Ishlinskii model shown in figure 7.9. Therefore, the equations (7.21) and (7.22) also represent the constitutive behavior of the Masing model corresponding to the input $u(t)$ and the output $F(t)$. Differentiating the equation (7.22) with respect to time yields

$$\dot{K}_i = \begin{cases} c_i \dot{u} & \text{if } |K_i| < H_i \text{ or } |K_i| = H_i \wedge \text{sgn}(\dot{u}K_i) \leq 0, \\ 0 & \text{otherwise.} \end{cases} \tag{7.52}$$

The evolution equation for the output force K_i acting on the i-th elasto-slide element $(i = 1, 2, \ldots, n)$ can be summarized as [88]

$$\dot{K}_i = \frac{1}{2} c_i \dot{u} \left\{1 - \text{sgn}(K_i^2 - H_i^2) - \text{sgn}(\dot{u}K_i)\left[1 + \text{sgn}(K_i^2 - H_i^2)\right]\right\}. \tag{7.53}$$

Due to the presence of the sign function, it is difficult to integrate this equation numerically. Indeed, the right-hand side of (7.53) is discontinuous and the exact value zero of the function $\text{sgn}(K_i^2 - H_i^2)$ when $|K_i| = H_i$ can hardly be obtained. To reduce these numerical difficulties, the following approximation is introduced [65]

$$\text{sgn}(K_i^2 - H_i^2) \approx \left|\frac{K_i}{H_i}\right|^m - 1 \quad \text{for} \quad |K_i| \leq H_i, \quad m \in \mathbb{R} \wedge m > 1. \tag{7.54}$$

Figure 7.12 illustrates this approximation for several values of the exponent m. The equation (7.53) then becomes

$$\dot{K}_i = c_i \dot{u} \left\{1 - \frac{1}{2}\left[1 + \text{sgn}(\dot{u}K_i)\right]\left|\frac{K_i}{H_i}\right|^m\right\} \quad (i = 1, 2, \ldots, n). \tag{7.55}$$

It is evident that the remaining sign function in (7.55) results in no discontinuity on the right-hand side, since $|K_i/H_i|^m$ or the whole right-hand side will vanish at the switching point of the function $\text{sgn}(\dot{u}K_i)$.

Chapter 7. Nonlinear quasi-static modeling

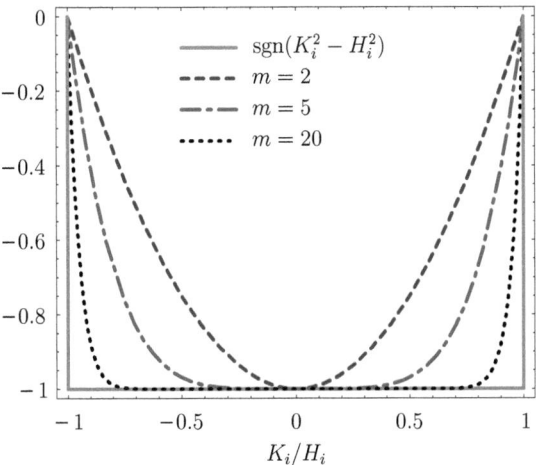

Figure 7.12: Comparision of the function $\operatorname{sgn}(K_i^2 - H_i^2)$ with its approximation $|K_i/H_i|^m - 1$.

7.1.4 Bouc-Wen model

The last hysteresis model used in this work is the Bouc-Wen model describing the dependence of the output $F(t)$ on the input $u(t)$, for instance the relation between forces and given displacements. This model was first proposed in [17, 18] where the force is described by a memory function in form of the Riemann-Stieltjes integral

$$K(t) = \int_0^t \mu[x(t) - x(\tau)] \, u' \, d\tau, \qquad (7.56)$$

with $()' = d/d\tau$, the derivative u' is assumed to be continuous and $x(t)$ is the total variation of u [20]

$$x(t) = \int_0^t \left|\frac{du}{d\tau}\right| du(\tau) \quad \Leftrightarrow \quad \dot{x} = |\dot{u}| \quad \text{with} \quad x(0) = 0. \qquad (7.57)$$

Applying the general form of the Leibniz integral rule

$$\dot{G}(t) = \frac{d}{dt}\int_{a(t)}^{b(t)} g(t,\tau) \, d\tau = g[t,b(t)]\frac{db}{dt} - g[t,a(t)]\frac{da}{dt} + \int_{a(t)}^{b(t)} \frac{\partial}{\partial t}g(t,\tau) \, d\tau \qquad (7.58)$$

to differentiate under the integral sign the function (7.56) with respect to t yields

$$\dot{K}(t) = \int_0^t \dot{\mu}[x(t) - x(\tau)] \, u' \, d\tau + \mu(0)\,\dot{u}(t). \qquad (7.59)$$

In the special case of an exponential function

$$\mu(x) = \alpha e^{-\gamma x} \quad \text{with} \quad \alpha, \gamma > 0, \qquad (7.60)$$

an evolution equation for the hysteresis force can be derived as

$$\dot{K} = \dot{u}\{\alpha - \gamma\,\text{sgn}(\dot{u}K)\,|K|\}. \tag{7.61}$$

This evolution equation was then extended in [125] as

$$\dot{K} = \dot{u}\{\alpha - [\beta + \gamma\,\text{sgn}(\dot{u}K)]\,|K|^m\}, \quad m \geq 1. \tag{7.62}$$

The equation (7.62) can be considered as a general case of (7.55) and it describes behavior of some hysteresis element called Bouc-Wen element.

More generally the Bouc-Wen model can be represented by a paralellel combination of the above Bouc-Wen elements similar to the construction of the Prandtl-Ishlinskii or Masing model. Then the behavior of this hysteresis system is described by

$$F(t) = c_0\,u(t) + \sum_{i=1}^{n} K_i(t), \tag{7.63}$$

$$\dot{K}_i = \dot{u}\{\alpha_i - [\beta_i + \gamma_i\,\text{sgn}(\dot{u}K_i)]\,|K_i|^m\}, \quad m \geq 1, \tag{7.64}$$

where c_0, α_i, β_i, γ_i and m are constant parameters of the Bouc-Wen model.

To identify the parameters of the model several methods were introduced in [65] such as using a dynamic programming, simplex and gradient method or multiple shooting method. Many other methods can be found in the review paper [53]. The later results in the present work show that a linear spring coupled in parallel with one Bouc-Wen element is appropriate to describe the hysteretic behavior of piezoceramics. Then the total force in the equation (7.63) reduces to

$$F(t) = c_0\,u(t) + K(t), \tag{7.65}$$

with the evolution equation for the hysteresis force K is given in (7.62). Therefore, the parameters of this simple Bouc-Wen model can be determined by using the built-in function $FindFit$ of $Mathematica$ which finds a least squares fit of the experimental first-order reversal curves.

7.2 Modeling of piezoceramics with hysteresis

The four models described in section 7.1 are now applied to simulate the nonlinear hysteretic behavior of transversally polarized piezoceramics subjected to moderate quasi-static loads. Fitting the experimental results given in chapter 6, the parameters of each model can be determined.

The classical Preisach (CP) model has theoretically an infinite number of parameters, which are the values of the Preisach function $\mu(\alpha,\beta)$ at all points inside the limiting triangle S. A discretization of S will naturally restrict to a finite number of parameters. This number of the values of the discretized Preisach function depends on the number of the first-order reversal curves obtained in experiments. With a set of n first-order reversal curves the Preisach plane S can be divided into $n(n+1)/2$ cells, inside each of them the Preisach function has a constant value.

The Prandtl-Ishlinskii and the Masing models have the same construction, thus the parameters determined for the Prandtl-Ishlinskii model will be used for both of them. The last model used in this work is the simple Bouc-Wen model containing only one hysteresis element.

7.2.1 Piezoceramics under moderate electric field

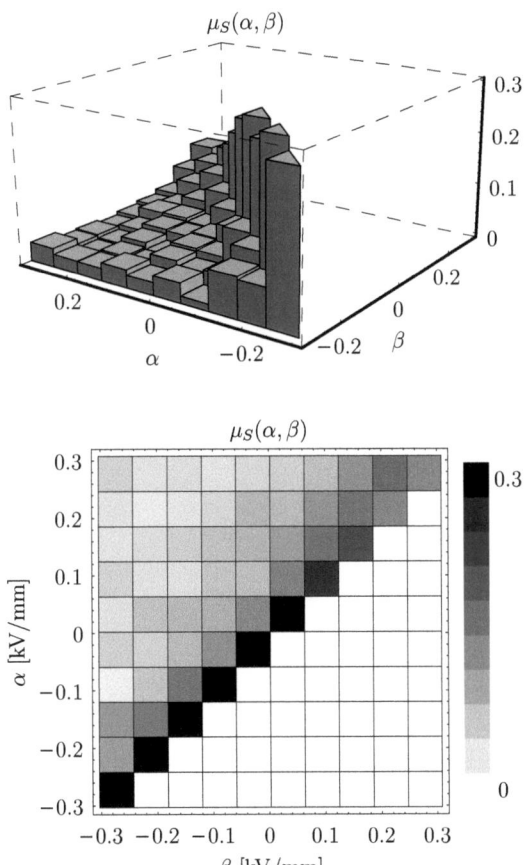

Figure 7.13: The identified Preisach function for the relation between applied electric field E_τ and longitudinal strain S_{xx} of PIC 255.

Following the identification procedure described for each hysteresis model in section 7.1, the corresponding parameters can be determined and then used to simulate the results from experiments. For the nonlinear hysteretic behavior of piezoceramics of the

84   Chapter 7. Nonlinear quasi-static modeling

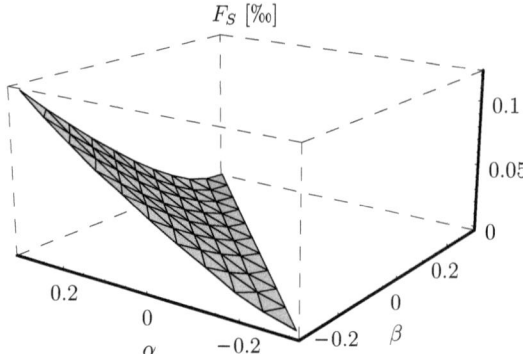

Figure 7.14: The identified Everett function for the relation between applied electric field E_z and longitudinal strain S_{xx} of PIC 255.

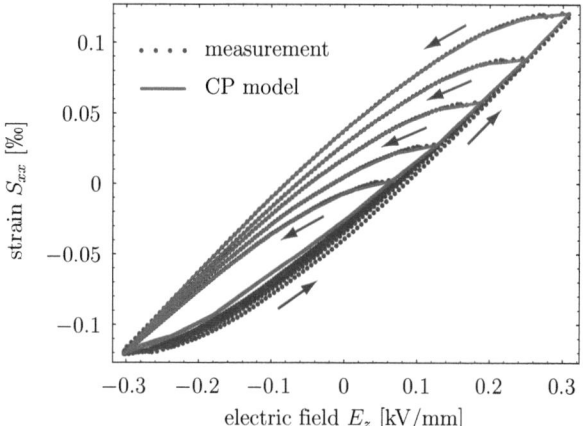

Figure 7.15: Comparision of the classical Preisach model with experiments for PIC 255: longitudinal strain S_{xx} vs. applied electric field E_z.

material PIC 255 under moderate quasi-static electric fields, both relations between the applied electric field E_z and the longitudinal strain S_{xx} as well as between E_z and the electric displacement density D_z are fitted. From the experimental first-order reversal curves, the Preisach and the Everett functions are first identified. Figure 7.13 shows the Preisach function $\mu_S(\alpha, \beta)$ for the strain output as a three-dimensional (3D) graphic together with a density plot. The discretization of the Preisach plane into $n = 10$ levels on each axes is based on the fact that a set of ten first-order reversal curves is taken from experiments. The corresponding Everett function $F_S(\alpha, \beta)$ is presented in figure 7.14. With these identified Preisach and Everett functions a very good

Chapter 7. Nonlinear quasi-static modeling   85

coincidence between theoretical and experimental results can be obtained as shown in figure 7.15, where only five reversal curves are plotted for ease of observation.

The same good result is also derived for the electric displacement output. The Preisach function $\mu_D(\alpha, \beta)$ and the Everett function $F_D(\alpha, \beta)$ are illustrated in figures 7.16 and 7.17 respectively. By comparision of the Everett functions in figures 7.14 and 7.17, it can be seen that the surface represented by $F_S(\alpha, \beta)$ has a larger curvature since the strain–electric field hysteresis loops open more than that for the electric displacement density. Using the identified functions leads again to an extreme coincidence between the Preisach model and experiments as shown in figure 7.18.

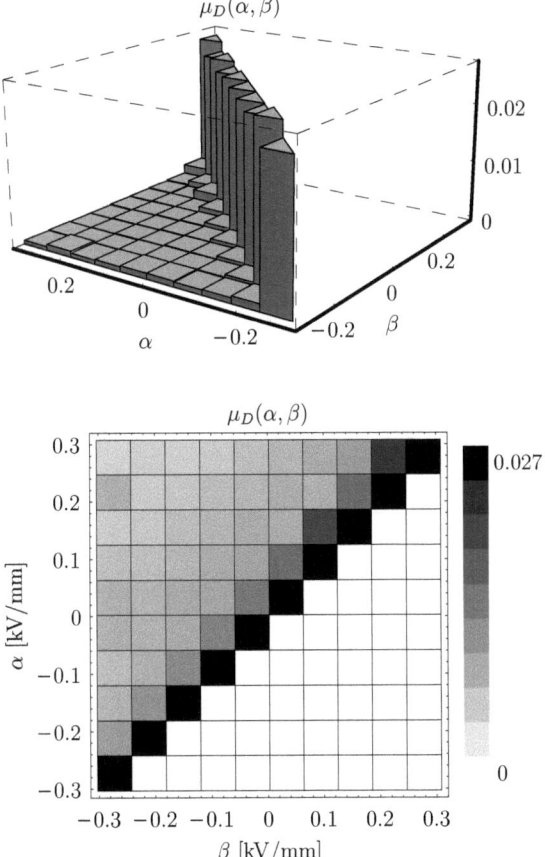

Figure 7.16: The identified Preisach function for the relation between applied electric field E_z and electric displacement density D_z of PIC 255.

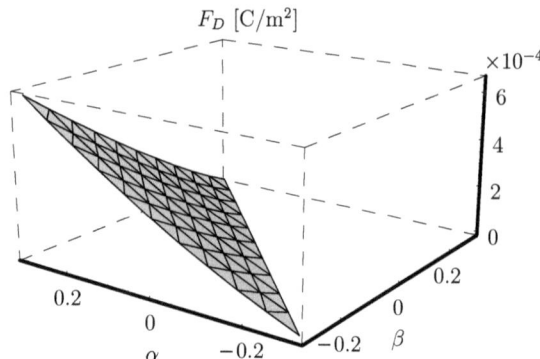

Figure 7.17: The identified Everett function for the relation between applied electric field E_z and electric displacement density D_z of PIC 255.

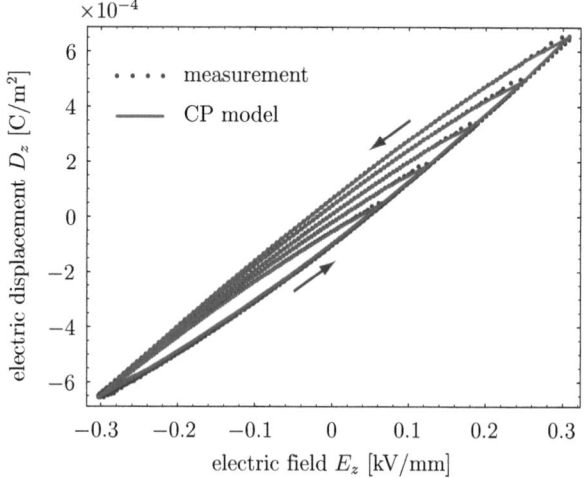

Figure 7.18: Comparision of the classical Preisach model with experiments for PIC 255: electric displacement density D_z vs. applied electric field E_z.

As a particular case of the classical Preisach model, the Prandtl-Ishlinskii model is also specified by its weight function and the respective Everett function. To model the dependence of the strain and the electric displacement density on the excitation electric field, the inverse PI model is applied because the ascending branches of the hysteresis loops are below the descending ones. Table 7.1 gives the identified parameters of the PI models containing $n = 5$ elasto-slide elements. The corresponding Preisach and Everett functions of the model for the strain response are presented in figures 7.19 and 7.20 respectively. It is noted after the equation (7.40) that an overline refers to the inverse

Chapter 7. Nonlinear quasi-static modeling   87

model. Here the characteristics of the Preisach and Everett functions of the PI model and its inverse mentioned in section 7.1.2 can visually be demonstrated: the Preisach function $\overline{\mu}_S(\overline{\alpha},\overline{\beta})$ consists of the line $\overline{\alpha} = \overline{\beta}$ and n lines parallel to this line; along the length of each line this function has a constant value which is directly related to the spring stiffnesses of the elasto-slide elements. The Everett function $\overline{F}_S(\overline{\alpha},\overline{\beta})$ represents a surface with the same form as that of the classical Preisach model. Figure 7.21 shows an acceptable agreement of the model with experimetal results. Theoretically the more elasto-slide elements, the better hysteresis curves resulting from the PI models. However, from experiments it is difficult to satisfy an essential property of the PI models that all hysteresis loops have a point of symmetry, thus non-physical negative values of spring stiffness or stiction force of a PI model with many elasto-slide elements may be derived. This problem can be overcome by using least square method with a constraint that all parameters of the PI model are positive, but it is not considered within this work.

i		0	1	2	3	4	5
S_{xx} [‰]	c_i	1.87461	0.10543	0.27529	0.33848	0.59349	2.45203
	H_i	∞	0.00789	0.01451	0.01131	0.01052	0.02174
D_z [C/m²]	c_i	385.281	37.8361	21.3268	35.0863	50.7523	78.3657
	H_i	∞	0.01799	0.00774	0.00907	0.00834	0.00644

Table 7.1: Identified parameters of the PI models with $n = 5$ for PIC 255.

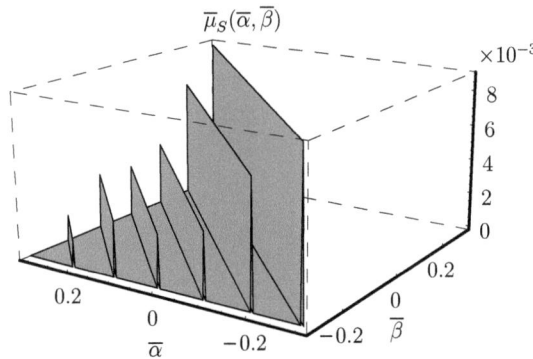

Figure 7.19: The identified Preisach function of the inverse PI model for the relation between applied electric field E_z and longitudinal strain S_{xx} of PIC 255.

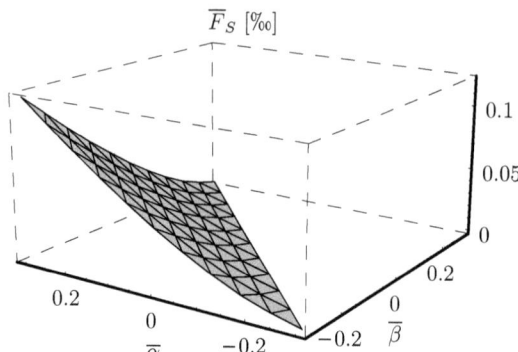

Figure 7.20: The identified Everett function of the inverse PI model for the relation between applied electric field E_z and longitudinal strain S_{xx} of PIC 255.

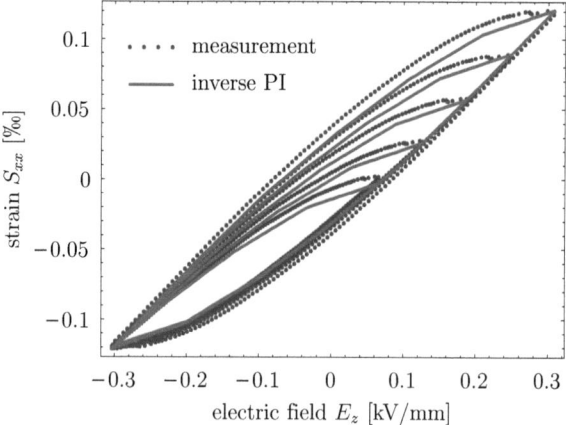

Figure 7.21: The inverse PI model with $n = 5$ compared to experiments for PIC 255: longitudinal strain S_{xx} vs. applied electric field E_z.

Chapter 7. Nonlinear quasi-static modeling

In contrast, a good coincidence between results of the inverse PI model and experiments can be obtained for the electric displacement response as shown in figure 7.24. The corresponding Preisach and Everett functions of the model are presented in figures 7.22 and 7.23 respectively. In this case the aforementioned characteristics of these functions are also verified.

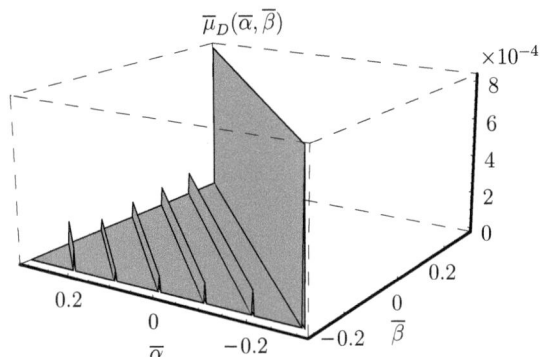

Figure 7.22: The identified Preisach function of the inverse PI model for the relation between applied electric field E_z and electric displacement density D_z of PIC 255.

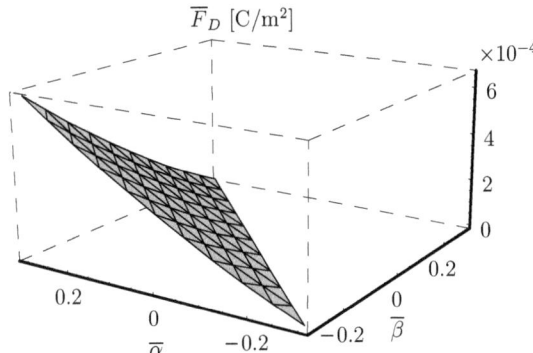

Figure 7.23: The identified Everett function of the inverse PI model for the relation between applied electric field E_z and electric displacement density D_z of PIC 255.

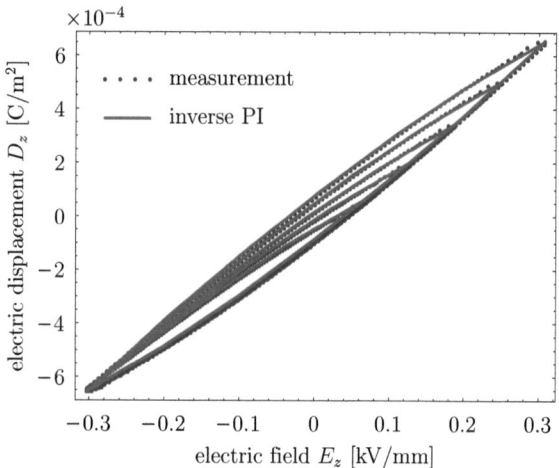

Figure 7.24: The inverse PI model with $n = 5$ compared to experiments for PIC 255: electric displacement density D_z vs. applied electric field E_z.

Next, introduce the identified parameters of the Prandtl-Ishlinskii model given in table 7.1 into the Masing model (7.55) with the exponent $m = 2$, a good accordance of theoretical results with experiments can be derived for both strain and electric displacement responses as shown in figures 7.25 and 7.26 respectively.

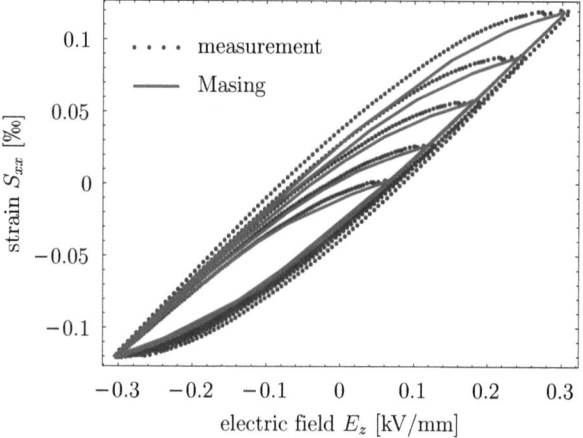

Figure 7.25: The Masing model with $n = 5$ compared to experiments for PIC 255: longitudinal strain S_{xx} vs. applied electric field E_z.

Chapter 7. Nonlinear quasi-static modeling

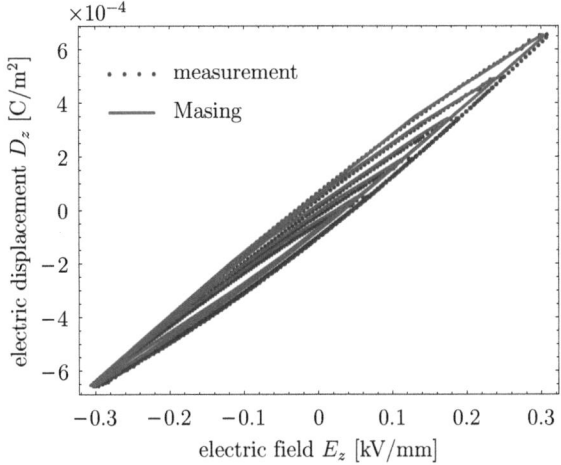

Figure 7.26: The Masing model with $n = 5$ compared to experiments for PIC 255: electric displacement density D_z vs. applied electric field E_z.

Finally the Bouc-Wen model (7.63) and (7.64) containing a single hysteresis element is used. Choosing the exponent $m = 2$ and fitting the first-order reversal curves, the parameters of the model can be determined as given in table 7.2. A good agreement between theoretical and experimental results for strain as well as electric displacement response are presented in figure 7.27 and 7.28 respectively.

Output	c_0	α	β	γ	m
S_{xx} [‰]	0.54	-0.32	126	292	2
D_z [C/m^2]	0.0026	-0.00075	-540	37000	2

Table 7.2: Identified parameters of the Bouc-Wen model with $n = 1$ for PIC 255.

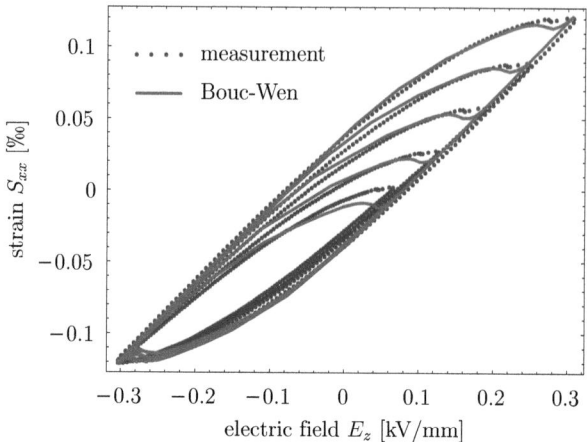

Figure 7.27: The Bouc-Wen model with $n = 1$ compared to experiments for PIC 255: longitudinal strain S_{xx} vs. applied electric field E_z.

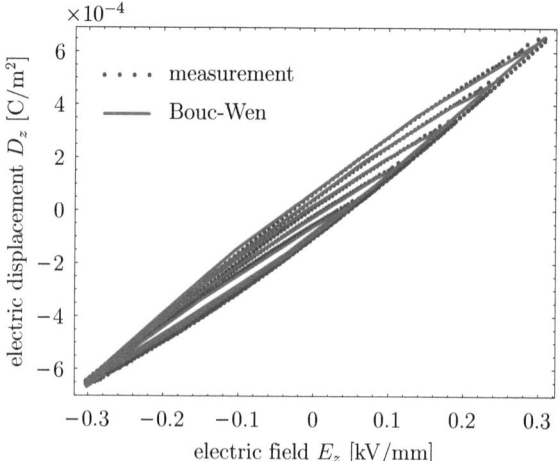

Figure 7.28: The Bouc-Wen model with $n = 1$ compared to experiments for PIC 255: electric displacement density D_z vs. applied electric field E_z.

7.2.2 Piezoceramics under moderate mechanical stress

Due to the difficulty of the tension tests described in section 6.2, only hysteresis stress–strain curves obtained from the compression tests are used. This means the stresses and strains are now negative unipolar or the hysteresis loops are defined in the third quadrant of the stress–strain plane. Therefore, the classical Preisach model should be modified so that the Preisach function can hold the physical property of a distribution function, i.e. it is a non-negative function.

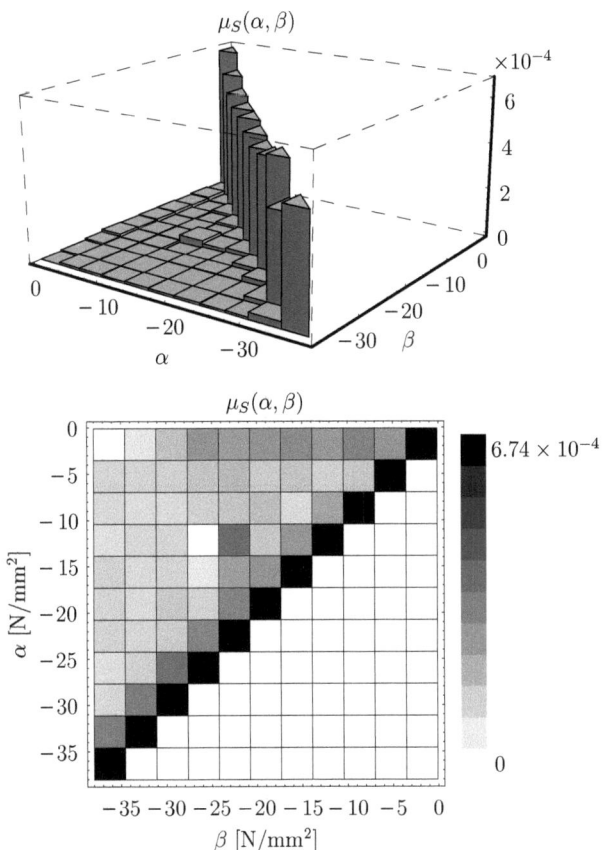

Figure 7.29: The identified Preisach function for the dependence of longitudinal strain S_{xx} upon applied stress T_{xx} of PIC 181.

The Preisach model is now defined as a weighted superposition of new elementary hysteresis operators $\hat{\gamma}_{\alpha\beta}$ of which the output values at "up" and "down" positions correspond to 0 and -1 instead of -1 and $+1$ as described in section 7.1.1. The Preisach function $\mu(\alpha,\beta)$ can be found by fitting the first-order increasing reversal curves (instead of the decreasing ones so far), each of which is formed when a monotonic decrease of the input from value above the "up" limiting α_0 to some value β is followed by a subsequent monotonic input increase [81]. If f_β denotes the output value at the reversal point $u(t) = \beta$ and $f_{\beta\alpha}$ is used for the output value on the respective first-order increasing reversal curve at $u(t) = \alpha$, the Everett function is modified for this case as

$$F(\alpha, \beta) = f_\beta - f_{\beta\alpha}. \tag{7.66}$$

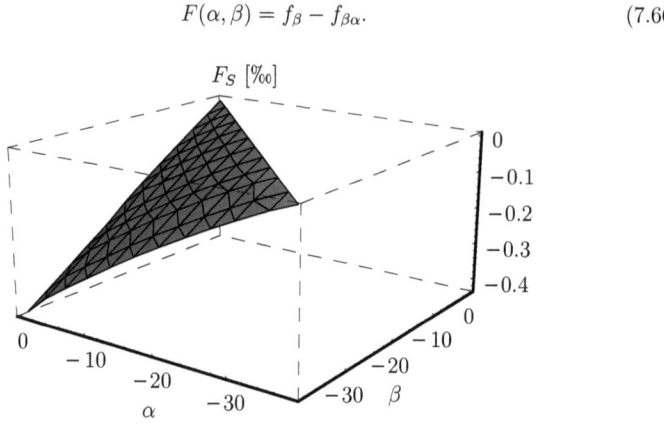

Figure 7.30: The identified Everett function for the dependence of longitudinal strain S_{xx} upon applied stress T_{xx} of PIC 181.

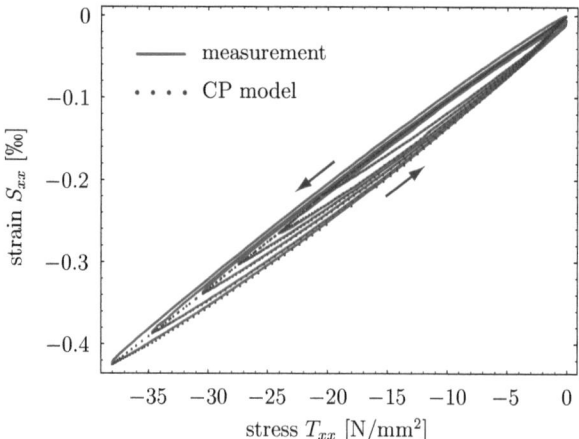

Figure 7.31: Comparision of the classical Preisach model with experiments for PIC 181: longitudinal strain S_{xx} vs. applied stress T_{xx}.

Using the modification of the elementary hysteresis operator and the Everett function the stress–strain relation of piezoceramics PIC 181 under moderate quasi-static

Chapter 7. Nonlinear quasi-static modeling   95

mechanical stresses can be simulated by a quite similar way to that described in section 7.1.1. First the Preisach and Everett functions of the model are identified by fitting the experimental first-order increasing reversal curves as shown in figures 7.29 and 7.30 respectively. In this case the stress T_{xx} is considered as the input, whereas the strain S_{xx} is the output. Figure 7.31 presents a good agreement between results of the Preisach model and those from experiments.

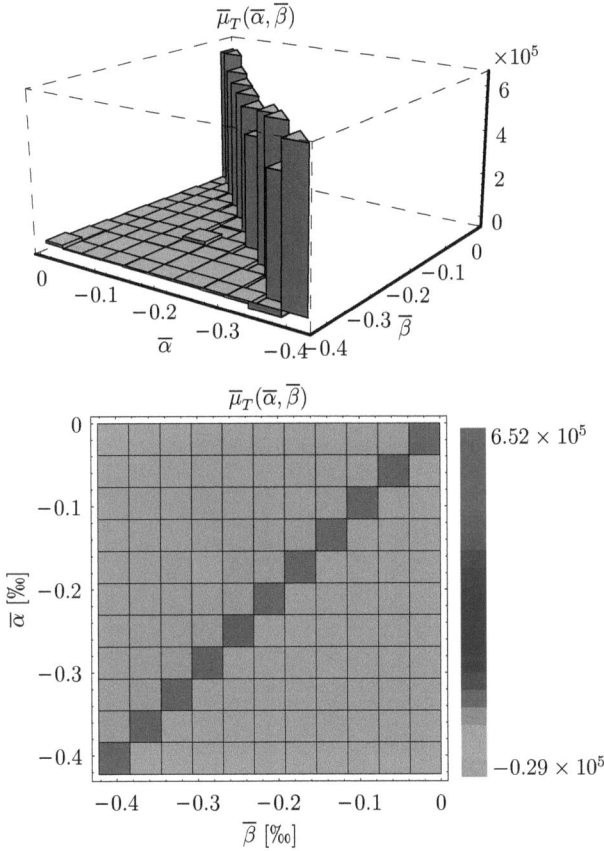

Figure 7.32: The identified Preisach function for the dependence of stress T_{xx} upon strain S_{xx} of PIC 181.

By contrast, if the dependence of the stress T_{xx} upon the strain S_{xx} is of interest, then the inverse Preisach model has to be used. The corresponding Preisach and Everett functions are determined by fitting the first-order reversal curves as demonstrated in figures 7.32 and 7.33 respectively. It can be seen that, the Preisach function of the inverse model has negative values. A good accordance of the modeling results with experiments can also be obtained as presented in figure 7.34.

96 Chapter 7. Nonlinear quasi-static modeling

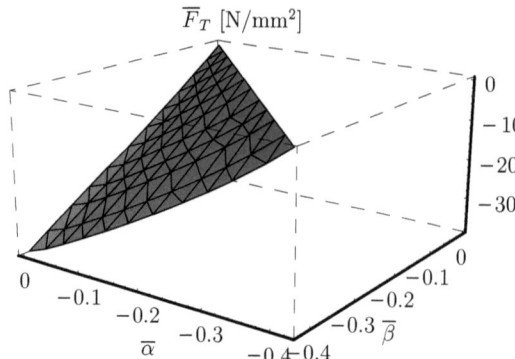

Figure 7.33: The identified Everett function for the dependence of stress T_{xx} upon strain S_{xx} of PIC 181.

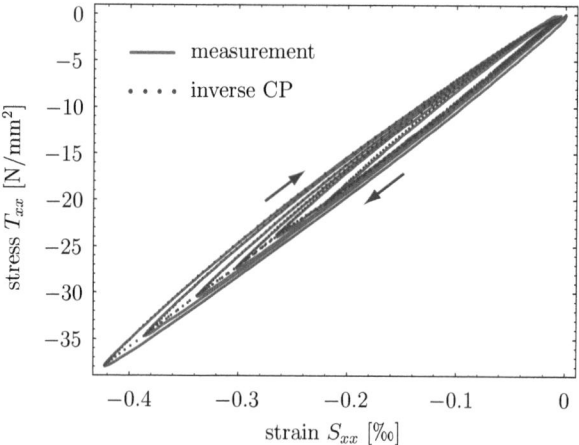

Figure 7.34: Comparision of the inverse Preisach model with experiments for PIC 181: applied stress T_{xx} vs. longitudinal strain S_{xx}.

In contrast to the Preisach models, the inverse Prandtl-Ishlinskii model is basically used for the dependence of the strain S_{xx} on the applied stress T_{xx}, whereas for the direct Prandtl-Ishlinskii model the strain is considered as the input and the stress is the output. The identified parameters of the PI model comprising $n = 5$ elasto-slide elements for each case are given in table 7.3 and 7.4. The Preisach and Everett functions corresponding to the inverse and direct PI model are presented in figures 7.35–7.36 and 7.38–7.39 respectively.

Chapter 7. Nonlinear quasi-static modeling

i	0	1	2	3	4	5
c_i [N/mm^2]	38735.0	4231.4	1093.8	1707.5	3479.2	4545.1
H_i [N/mm^2]	∞	1.295	0.26	0.295	0.388	0.254

Table 7.3: Identified parameters for the PI model with $n = 5$ for the dependence of strain S_{xx} upon stress T_{xx} of PIC 181.

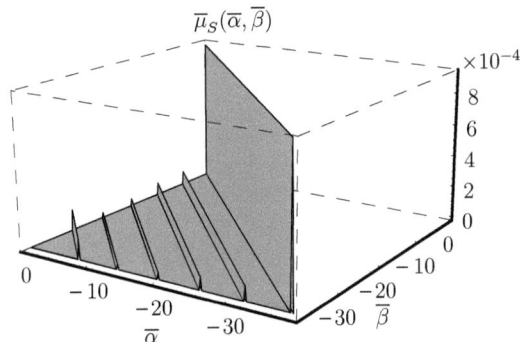

Figure 7.35: The identified Preisach function of the inverse PI model with $n = 5$ for the dependence of longitudinal strain S_{xx} upon applied stress T_{xx} of PIC 181.

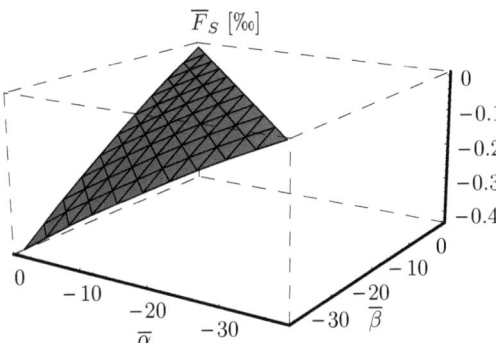

Figure 7.36: The identified Everett function of the inverse PI model with $n = 5$ for the dependence of longitudinal strain S_{xx} upon applied stress T_{xx} of PIC 181.

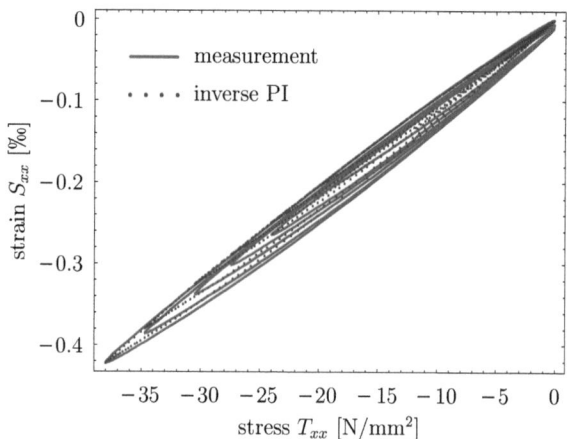

Figure 7.37: The inverse PI model with $n = 5$ compared to experiments for PIC 181: longitudinal strain S_{xx} vs. applied stress T_{xx}.

i	0	1	2	3	4	5
c_i [N/mm^2]	142539	22420	12184	14078	8373	19072
H_i [N/mm^2]	∞	3.699	1.584	1.408	0.544	0.668

Table 7.4: Identified parameters for the PI model with $n = 5$ for the dependence of stress T_{xx} upon strain S_{xx} of PIC 181.

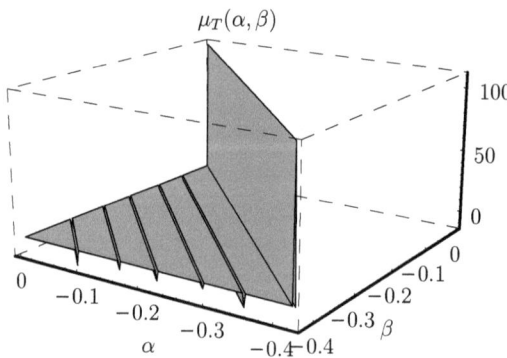

Figure 7.38: The identified Preisach function of the PI model with $n = 5$ for the dependence of applied stress T_{xx} upon longitudinal strain S_{xx} of PIC 181.

Chapter 7. Nonlinear quasi-static modeling 99

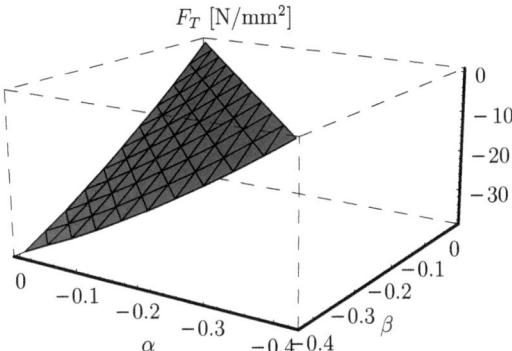

Figure 7.39: The identified Everett function of the PI model with $n = 5$ for the dependence of applied stress T_{xx} upon longitudinal strain S_{xx} of PIC 181.

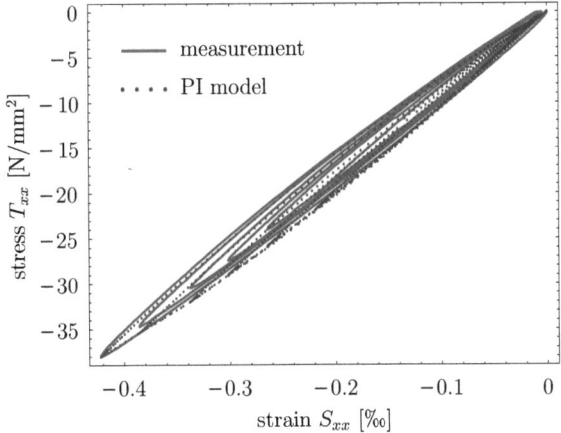

Figure 7.40: The PI model with $n = 5$ compared to experiments for PIC 181: applied stress T_{xx} vs. longitudinal strain S_{xx}.

Since the applied stress T_{xx} is now negative unipolar, the identified parameters of the PI model given in table 7.4 can no longer be introduced into the Masing model. In this case, parameters of the Masing model has to be determined by using the identification procedure for the PI model in the following way. First the experimental first-order reversal curves are shifted so that the zero point is the center of the major hystereis stress-strain loop, this changes the stress and the strain to become bipolar. Applying then the identification method for the PI model to fit these modified first-order reversal curves, the parameters c_i and H_i of the Masing model can be found as given in table 7.5. As a result of the offset, the identified parameters here are about half of those given in table 7.4 for the PI model with the same structure.

i	0	1	2	3	4	5
c_i [N/mm²]	70301.1	12178.6	6091.8	7038.9	4186.5	9535.9
H_i [N/mm²]	∞	2.01	0.792	0.704	0.272	0.334

Table 7.5: Identified parameters of the Masing model with $n = 5$ for PIC 181.

The stress–strain relation is now modeled by using the equations (7.21) and (7.55) with the exponent $m = 2$ as

$$T_{xx} = c_0 S_{xx} + \sum_{i=1}^{5}[K_i(t) - H_i], \tag{7.67}$$

$$\dot{K}_i = c_i \dot{S}_{xx} \left\{ 1 - \frac{1}{2}\left[1 + \mathrm{sgn}(\dot{S}_{xx} K_i)\right] \left|\frac{K_i}{H_i}\right|^2 \right\} \quad (i = 1, 2, 3, 4, 5). \tag{7.68}$$

A good agreement of theoretical and experimental results is derived as presented in figure 7.41.

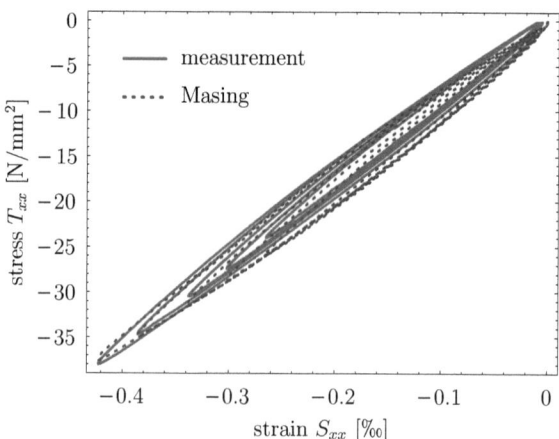

Figure 7.41: The Masing model with $n = 5$ compared to experiments for PIC 181: applied stress T_{xx} vs. longitudinal strain S_{xx}.

Last the Bouc-Wen model with a single hysteresis element is used to simulate the stress–strain hysteresis. Table 7.6 gives the parameters of the model determined by fitting the experimental first-order reversal curves. Introducing these parameters into the equations (7.63) and (7.64) results in a relatively good accordance of modeling results with experiments as demonstrated in figure 7.42.

$c_0 \left[\dfrac{\text{N}}{\text{mm}^2}\right]$	$\alpha \left[\dfrac{\text{N}}{\text{mm}^2}\right]$	$\beta \left[\dfrac{\text{mm}^2}{\text{N}}\right]$	$\gamma \left[\dfrac{\text{mm}^2}{\text{N}}\right]$	m
82491.6	13238.3	−3030.93	4.85461	2

Table 7.6: Identified parameters of the Bouc-Wen model with $n = 1$ for PIC 181.

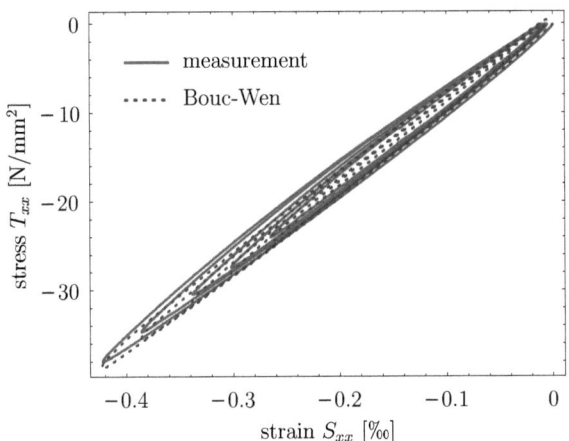

Figure 7.42: The Bouc-Wen model with $n = 1$ compared to experiments for PIC 181: applied stress T_{xx} vs. longitudinal strain S_{xx}.

7.2.3 Comparision of hysteresis models

Comparing the results of the four hysteresis models presented in sections 7.2.1 and 7.2.2, it is obvious that the classical Preisach model is the best giving an extreme coincidence between the theoretical and experimental first-order reversal curves. As particular cases of the Preisach model, the Prandtl-Ishlinskii model and its inverse result also in an agreement of the modelings with experiments. Since the Preisach and Everett functions of the PI models are analytic expressions, it is easy to calculate their values over the Preisach plane. However, the symmetry of these functions restricts the flexibility of the PI models compared to the Preisach model considering the shape of the hysteresis curves. For both the Preisach and PI models a history of the input is required as a set of its dominant extrema, but all evaluations are only arithmetic. By contrast, using the Masing and Bouc-Wen models only differential evolution equations with internal variables have to numerically be solved without the requirement for the input history. Therefore, they can easily be combined with the dynamic modeling to describe the nonlinear behavior of piezoceramic oscillators as shown in next chapter. Nevertheless, errors of these models are more significant than those of the others. The results of the Masing and Bouc-Wen models are sensitive to the accuracy of measurements including the initial conditions.

Chapter 8

Combination of nonlinear modelings

As mentioned at the end of section 5.5, it is expected that the nonlinear dynamic behavior of piezoceramics subjected to weak electric fields can be explained by using the hysteresis stress–strain nonlinearities. First, instead of the nonlinear model described in chapter 5, the differential hysteresis models, namely the Masing and Bouc-Wen models are integrated into a linear conservative vibration system. The parameters of these models determined from the compression tests as given in section 7.2.2 can be applied to describe harmonic vibration behavior of piezoceramics because the hysteresis stresses represented by the models depend only on the rate of the strains. Moreover, the derivative of the hysteresis stress with respect to the strain dK/dS_{xx} depends only on the sign of the strain rate. This means the hysteresis stress–strain relations described by the Masing and Bouc-Wen models are independent of frequency. Another method is introducing the mechanical parameters determined from the stress–strain hysteresis into the dynamic nonlinear model.

8.1 Dynamic modeling with Masing hysteresis

In this section, a polarized piezoceramic rod subjected to weak electric fields with respect to the inverse 31-effect is considered as a vibration system, whose nonlinear quasi-static mechanical behavior is represented by a Masing model containing a single elasto-slide element for a simple case. Using Hamilton's principle, the equations of motion can be derived. Here the electric enthalpy density for the linear modeling given in the equation (4.10) is used

$$H = \frac{1}{2} E_0^{(0)} S_{xx}^2 - \gamma_0 S_{xx} E_z - \frac{1}{2} \nu_0 E_z^2. \tag{8.1}$$

The linear conservative constitutive law for the stress T_{xx} is then extended by a hysteresis term K as

$$T_{xx} = E_0^{(0)} S_{xx} - \gamma_0 E_z + K, \tag{8.2}$$

$$\dot{K} = E_1^{(0)} \dot{S}_{xx} \left\{ 1 - \frac{1}{2} \left[1 + \text{sgn}(\dot{S}_{xx} K) \right] \left| \frac{K}{H_1} \right|^2 \right\}. \tag{8.3}$$

104 Chapter 8. Combination of nonlinear modelings

The parameters of this Masing model determined from the stress–strain relations are given in table 8.1. For a single elasto-slide element, the corresponding well simulated results compared to experiments can even be obtained as shown in figure 8.1.

The kinetic energy density T and the virtual work δW can be expressed as

$$T = \frac{1}{2}\rho \dot{u}^2(x,t) \quad \text{and} \quad \delta W = -A_p \int_{-\frac{l}{2}}^{\frac{l}{2}} K\,\delta u'\,\mathrm{d}x, \tag{8.4}$$

where A_p is the cross-section area of the piezoceramic rod.

i	0	1
$E_i^{(0)}$ [N/mm^2]	79057.7	22611.4
H_i [N/mm^2]	∞	2.261

Table 8.1: Identified parameters of the Masing model with $n = 1$ for PIC 181.

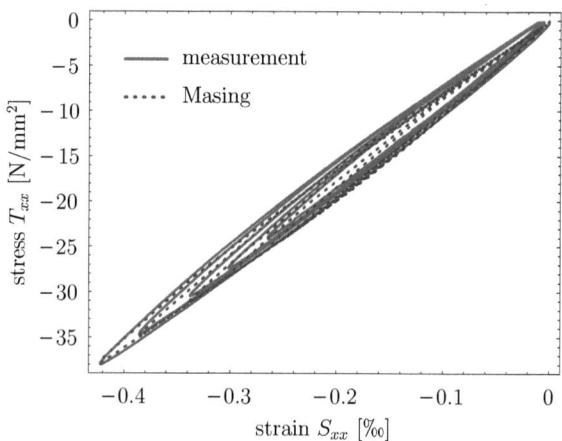

Figure 8.1: The Masing model with $n = 1$ compared to experiments for PIC 181: applied stress T_{xx} vs. longitudinal strain S_{xx}.

It is pointed out in section 5.2 that for the excitation near to the k-th eigenfrequency, the displacement response in the 1-direction of piezoceramic rod can be approximated by the single-mode Ritz ansatz as

$$u(x,t) = U_k(x)\,p(t), \tag{8.5}$$

where $U_k(x)$ is the sinusoidal eigenfunction corresponding to k-th eigenfrequency of the linear system as given in the equation (4.24). Substituting this function into Hamilton's principle (2.11) and carrying out the variations leads to the following nonlinear

Chapter 8. Combination of nonlinear modelings

equations of motion

$$m\ddot{p} + c^{(1)} p = f^{(1)} \cos \Omega t - \int_{-\frac{l}{2}}^{\frac{l}{2}} K U'_k(x) \, dx, \quad (8.6)$$

$$\dot{K} = U'_k(x) E_1^{(0)} \dot{p} \left\{ 1 - \frac{1}{2} \left[1 + \text{sgn}(\dot{p} K) \right] \left| \frac{K}{H_1} \right|^2 \right\}, \quad (8.7)$$

with keeping in mind that $S_{xx} = u'$, $U'_k(x) \geq 0$ for $-l/2 \leq x \leq l/2$ and

$$m = \rho \int_{-\frac{l}{2}}^{\frac{l}{2}} U_k^2(x) \, dx, \quad c^{(1)} = E_0^{(0)} \int_{-\frac{l}{2}}^{\frac{l}{2}} U'^2_k(x) \, dx, \quad f^{(1)} = \gamma_0 \frac{U_0}{h} \int_{-\frac{l}{2}}^{\frac{l}{2}} U'_k(x) \, dx.$$

Assuming that the hysteresis stress K is independent of position x, the above equations of motion can be expressed as

$$m\ddot{p} + c^{(1)} p = f^{(1)} \cos \Omega t - i_d K, \quad (8.8)$$

$$\dot{K} = \frac{i_d}{l} E_1^{(0)} \dot{p} \left\{ 1 - \frac{1}{2} \left[1 + \text{sgn}(\dot{p} K) \right] \left| \frac{K}{H_1} \right|^2 \right\}, \quad (8.9)$$

where

$$i_d = \int_{-\frac{l}{2}}^{\frac{l}{2}} U'_k(x) \, dx.$$

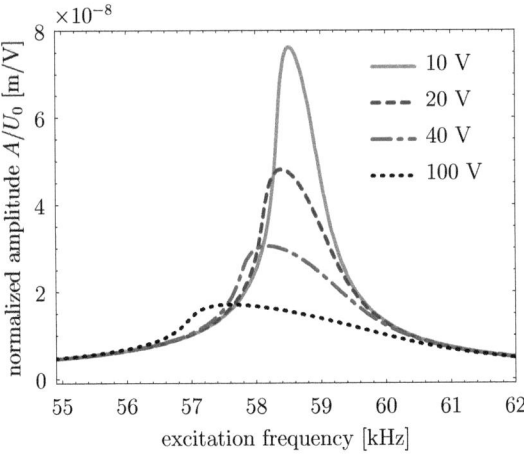

Figure 8.2: Normalized amplitude response with Masing hysteresis for PIC 181.

Using the mass density of PIC 181 given by the manufacturer $\rho = 7850$ kg/m^3, the piezoelectric constant $d_{31} = -1.11 \times 10^{-10}$ m/V determined in section 5.4, the hysteretic parameters given in table 8.1 and giving the excitation voltage $U(t) = U_0 \cos \Omega t$, the stationary amplitude–frequency response can be found by solving numerically the equations (8.8) and (8.9). Some results of such a calculation are presented in figure 8.2, where the displacement amplitude is normalized by the amplitude of the corresponding excitation voltage. It is very interesting that the present model gives rise to the nonlinear effects of Duffing-type observed from dynamic experiments including decrease of the first resonance frequency and the normalized displacement amplitude with increasing excitation voltage. Nevertheless, there is no jump of the displacement amplitude although the low damping material PIC 181 is considered. In addition, the resonance amplitude and frequency resulting from the model are different from those of experiments. Figure 8.3 shows a comparison of theoretical with experimental results for the excitation voltage at 40 V. It is obvious that despite having the same order of magnitude, the modeling resonant amplitude is only about 40% of the experimental one. Due to the superposition of two elastic moduli of the chosen Masing model, the modeling resonant frequency is greater than that from experiment, which corresponds to the linear modulus determined in section 5.4.

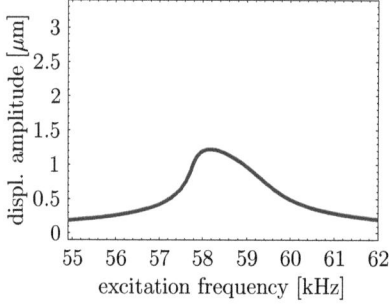
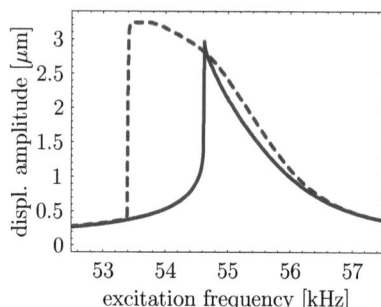

Figure 8.3: Displacement amplitude response of PIC 181 at 40 V. Left: system with Masing hysteresis. Right: experimental sweep up (solid line) and sweep down (dashed line) of the excitation frequency.

8.2 Dynamic modeling with Bouc-Wen hysteresis

As a further try to integrate a hysteresis model into the dynamic one, the Bouc-Wen model is used here. Following the same way applied to the Masing model, the linear conservative constitutive law of piezoceramics is extended by a hysteresis term K as

$$T_{xx} = E^{(0)} S_{xx} - \gamma_0 E_z + K, \tag{8.10}$$

$$\dot{K} = \dot{S}_{xx} \left\{ \alpha - \left[\beta + \gamma \operatorname{sgn}(\dot{S}_{xx} K) \right] K^2 \right\}, \tag{8.11}$$

with the parameters of the Bouc-Wen model are given in table 7.6 and $E^{(0)} = c_0$. Applying again Hamilton's principle with the single-mode Ritz ansatz (8.5) and the assumption that $K = K(t)$ results in nonlinear equations of motion

$$m\ddot{p} + c^{(1)} p = f^{(1)} \cos \Omega t - i_d K, \tag{8.12}$$

$$\dot{K} = \frac{i_d}{l} \dot{p} \left\{ \alpha - \left[\beta + \gamma \operatorname{sgn}(\dot{p} K) \right] K^2 \right\}, \tag{8.13}$$

Chapter 8. Combination of nonlinear modelings

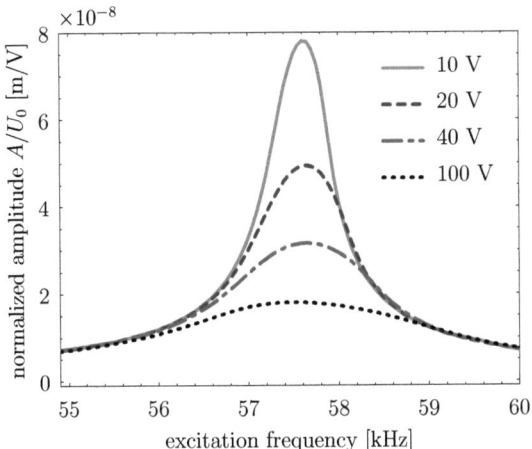

Figure 8.4: Normalized amplitude response with Bouc-Wen hysteresis for PIC 181.

where

$$m = \rho \int_{-\frac{l}{2}}^{\frac{l}{2}} U_k^2(x)\,dx, \ c^{(1)} = E^{(0)} \int_{-\frac{l}{2}}^{\frac{l}{2}} U_k'^2(x)\,dx, \ f^{(1)} = \gamma_0 \frac{U_0}{h} i_d, \ i_d = \int_{-\frac{l}{2}}^{\frac{l}{2}} U_k'(x)\,dx.$$

Similar to the modeling with Masing hysteresis, solving numerically the equations (8.12) and (8.13) for the stationary solution, the normalized amplitude response of piezoceramic rod excited by the harmonic voltage near resonance is obtained. Figure 8.4 shows such amplitude responses at various excitation voltages. These results are comparable in both resonance amplitude and frequency to those for the system with Masing hysteresis presented in figure 8.1. Decrease of the normalized displacement amplitude with increasing excitation voltage can also be observed, but the behavior of Duffing oscillator with hardening cubic stiffness is revealed and the differences in resonance amplitude and frequency between the model and experiments are remaining. This may result from the error of parameters identification. The use of different samples (with the same material and dimensions) in quasi-static and dynamic experiments can also account for the deviation of the resonance frequency. In further investigations, the quasi-static and dynamic experiments must successively be performed on the same sample, but it is difficult to do with relatively small samples (30 × 3 × 2 mm³) as used in this work. The quasi-static experiment should be done first because the strain gauges have to be glued on the clean surfaces of piezoceramics. Besides, epoxy is also used to prevent the slight slip of the sample in the clamps. Therefore, for the next dynamic experiment it is difficult to recover the clean initial state of the piezoceramics. Due to technical restrictions of the author in experiments, the dynamic behavior of piezoceramics can only be obtained by using laser vibrometer instead of the strain gauges. In addition, the assumption of the independence of the hysteresis force upon position x has a specified role in the deviation of the modelings from the experiments.

8.3 Dynamic modeling with variable mechanical parameters

In this section, another possibility to account for the nonlinear dynamic effects occuring in experiments using the hysteretic stress–strain relation will be presented. From the results in section 7.2.2, it can be seen that this hysteretic behavior has been well simulated by the classical Preisach model. Keep in mind that this model has basically congruency property, i.e. hysteresis loops resulting from back-and-forth variations of inputs between the same two consecutive extremum values are congruent [81]. This property can be verified and approximately extended by the result of a compression test shown in figure 8.5, where three almost congruent minor hysteresis loops are produced as the stress varies back-and-forth between consecutive extremum values whose difference are nearly the same. This extension is assumed to be valid for thin hysteresis nonlinearity, corresponding to that the Preisach function concentrates in adjacent region of the line $\alpha = \beta$. Examples of such Preisach functions are illustrated in figure 7.16 or 7.29.

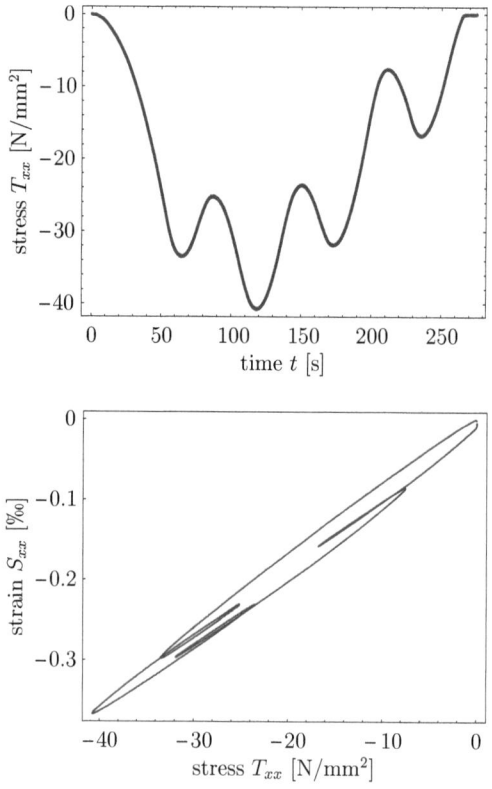

Figure 8.5: Congruent stress–strain hysteresis loops for PIC 181.

Chapter 8. Combination of nonlinear modelings 109

With the above important assumption, the experimental stress–strain hysteresis curves are academically shifted so that the zero point is the center of each hysteresis loop. Figure 8.6 shows the results of this academic step. The modified curves are now considered as dynamic responses of piezoceramics subjected to weak electric field, which is generated by the applied voltage $U(t) = U_0 \cos \Omega t$, close to the first resonance. This is based on the fact that the maximum value of the strains considered here is of the same order as that obtained from the dynamic experiments. Therefore, the longitudinal displacement of the piezoceramic rod can be expressed by using the single-mode Ritz ansatz containing the first eigenfunction and the harmonic time function with excitation frequency as

$$u(x,t) = A \sin \frac{\pi x}{l} \cos \Omega t. \tag{8.14}$$

Substituting the solution (8.14) into (2.5) yields the dynamic longitudinal strain and its derivative with respect to time as

$$S_{xx} = \widehat{S}(x) \cos \Omega t, \tag{8.15}$$
$$\dot{S}_{xx} = -\Omega \widehat{S}(x) \sin \Omega t, \tag{8.16}$$

where

$$\widehat{S}(x) = A \frac{\pi}{l} \cos \frac{\pi x}{l}. \tag{8.17}$$

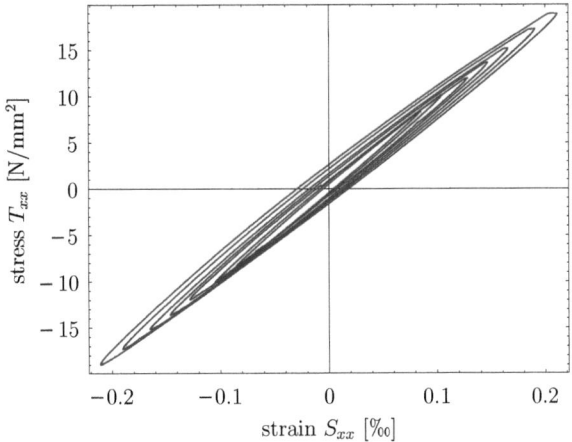

Figure 8.6: Shifted stress–strain hysteresis loops for PIC 181.

To identify the parameters of piezoceramics, some constitutive piezoelectric relation will be chosen, for example

$$T_{xx} = E^{(0)} S_{xx} - \gamma_0 E_z + E_d^{(0)} \dot{S}_{xx} - \gamma_{0d} \dot{E}_z. \tag{8.18}$$

In the quasi-static tension and compression tests, two electrodes of the piezoelectric samples are short-circuited, so the electric field E_z vanishes. This eliminates the influence of all piezoelectric coupling parameters. Giving the excitation frequency and fitting the theoretical stress–strain curves with the modified hysteresis loops shown in figure 8.6, the mechanical parameters of piezoceramics e.g. $E^{(0)}$ and $E_d^{(0)}$ can be determined. These parameters are no longer constant as so far but they depend on the strain amplitude $\widehat{S}$, thus on the displacement amplitude A and on the position x, as well as on the excitation frequency. However, close to the first resonance it can be assumed that the mechanical parameters are approximately independent of frequency. This is verified by the calculations taking account of the frequency dependence of the parameters, from which the same results are obtained. Introducing the identified mechanical parameters into the nonlinear dynamic model described in chapter 5 with given displacement amplitude and excitation frequency, the necessary excitation voltage can be calculated. Therefore, a sweep of displacement amplitude together with a sweep of excitation frequency will result in a set of amplitude–frequency responses corresponding to various amplitudes of the applied voltage. In the following several mechanical constitutive equations are considered, leading to respective amplitude responses as contour plots in comparision with dynamic experimental results. The excitation frequency used to determine the mechanical parameters is specified by the first resonance frequency of piezoceramic rod of PIC 181, i.e. about 55.8 kHz, as shown in figure 5.10.

Starting from the simplest case, the linear conservative constitutive equation is

$$T_{xx} = E^{(0)} S_{xx}. \tag{8.19}$$

The identified elastic modulus of PIC 181 in dependence on strain amplitude is shown in figure 8.7, where each point results from a hysteresis stress–strain loop and the solid line is the result of fitting the set of these points with a linear function. From this linear relation it can be inferred that the conservative modeling possesses the quadratic nonlinearity. This accounts for nonlinear effects shown in figure 8.8 which is a contour plot of the displacement amplitude response for excitation voltage up to 100 V. The resonance amplitudes go to infinity since there is no damping in this case.

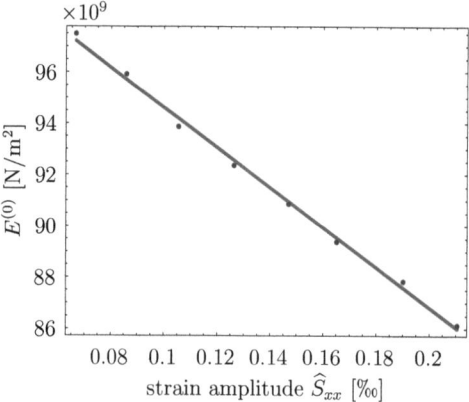

Figure 8.7: Elastic modulus of PIC 181.

Chapter 8. Combination of nonlinear modelings

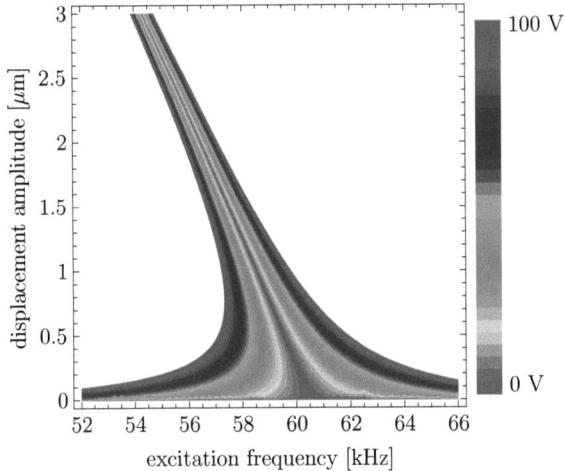

Figure 8.8: Amplitude response with linear conservative parameter for PIC 181.

Introducing a linear damping term into the linear conservative constitutive equation leads to
$$T_{xx} = E^{(0)} S_{xx} + E_d^{(0)} \dot{S}_{xx}. \tag{8.20}$$
Then the stress–strain hysteresis can well be simulated as presented in figure 8.10. The results for identified elastic modulus and linear damping of the material can be seen in figure 8.9. Both of them are fitted with linear functions which are plotted with the solid lines. The corresponding displacement amplitude response is shown in figure 8.11. Due to the presence of the higher linear damping in comparision with that identified from the dynamic experiments given in table 5.2, the jump phenomena and multiple stable amplitude responses can no longer be observed.

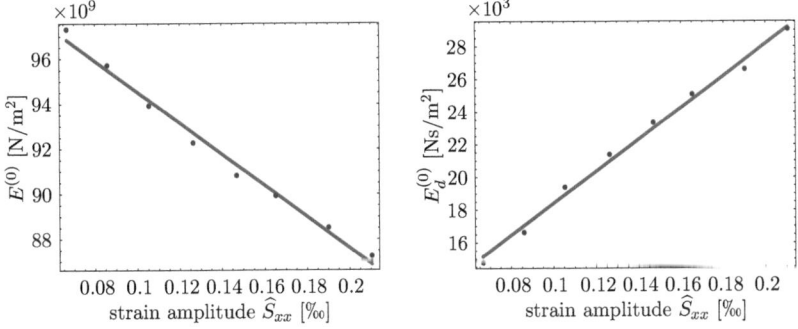

Figure 8.9: Linear parameters of PIC 181.

112   Chapter 8. Combination of nonlinear modelings

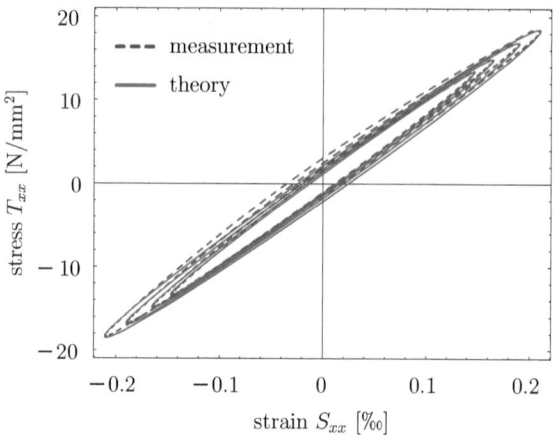

Figure 8.10: Comparision of the stress–strain hysteresis loops for PIC 181.

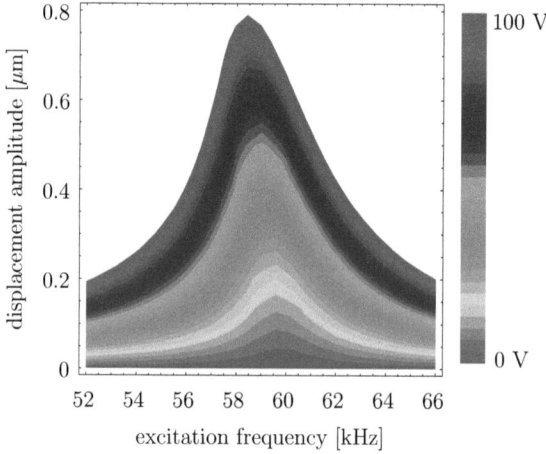

Figure 8.11: Amplitude response with linear parameters for PIC 181.

We consider now a nonlinear constitutive equation consisting of both conservative and nonconservative quadratic and cubic terms as

$$T_{xx} = E^{(0)} S_{xx} + E_d^{(0)} \dot{S}_{xx} + E^{(1)} S_{xx}^2 + E_d^{(1)} (\dot{S_{xx}^2}) + E^{(2)} S_{xx}^3 + E_d^{(2)} (\dot{S_{xx}^3}). \quad (8.21)$$

As demonstrated in figure 8.12, a very good coincidence between the theoretical and experimental hysteresis curves is obtained. Figure 8.14 gives the results of fitting the identified parameters with compatible functions. Linear functions are used for

Chapter 8. Combination of nonlinear modelings   113

$E^{(0)}$, $E_d^{(0)}$, $E_d^{(1)}$ and $E_d^{(2)}$ whereas a quadratic function for $E^{(1)}$ and a cubic one for $E^{(2)}$. Introducing these parameters into the nonlinear dynamic model results in the displacement amplitude responses plotted in figure 8.13.

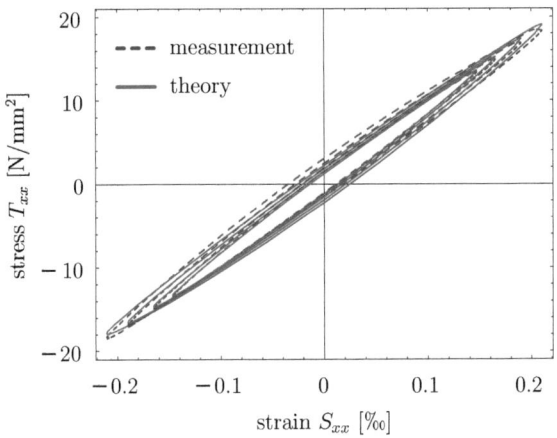

Figure 8.12: Comparision of the stress–strain hysteresis loops for PIC 181.

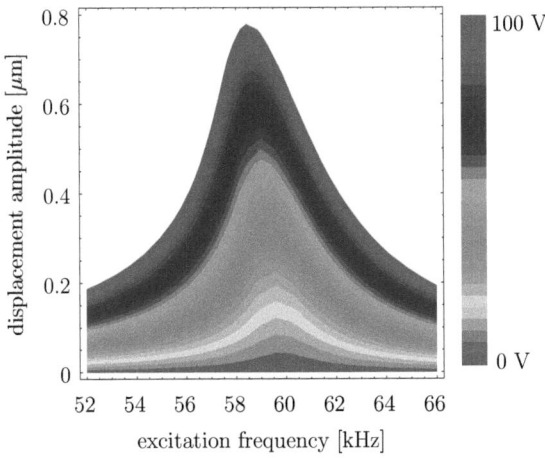

Figure 8.13: Amplitude response with mechanical nonlinearities for PIC 181.

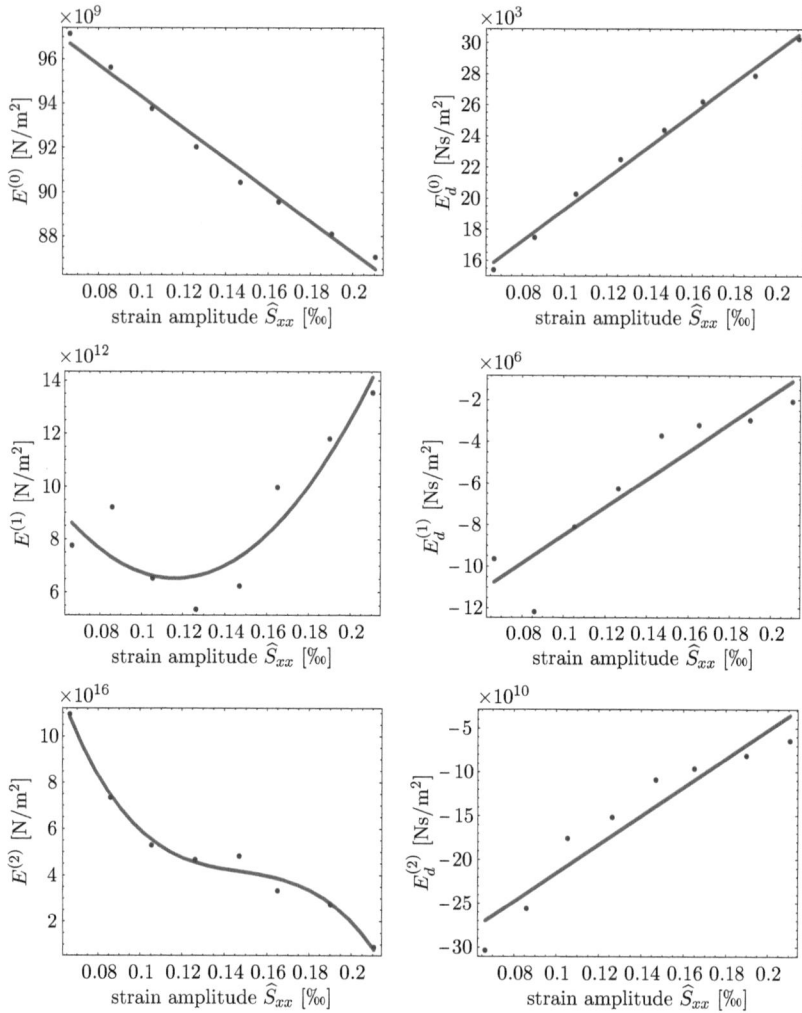

Figure 8.14: Nonlinear mechanical parameters of PIC 181.

In case the nonlinear constitutive equation contains no quadratic terms as

$$T_{xx} = E^{(0)} S_{xx} + E_d^{(0)} \dot{S}_{xx} + E^{(2)} S_{xx}^3 + E_d^{(2)} (\dot{S_{xx}^3}), \tag{8.22}$$

a good agreement of the simulated stress–strain hysteresis relation with experiments is also derived as shown in figure 8.15. Fitting the identified parameters can be done in the same way as described above, namely for $E^{(0)}$, $E_d^{(0)}$ and $E_d^{(2)}$ linear functions are used and a cubic function for $E^{(2)}$. Figure 8.16 shows the results of such a process and figure 8.17 presents the corresponding results for the displacement amplitude responses.

Chapter 8. Combination of nonlinear modelings

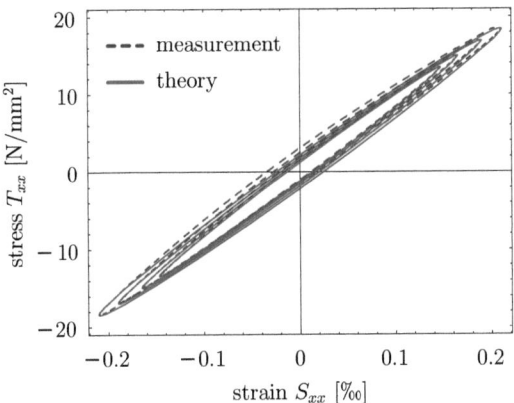

Figure 8.15: Comparision of the stress–strain hysteresis loops for PIC 181.

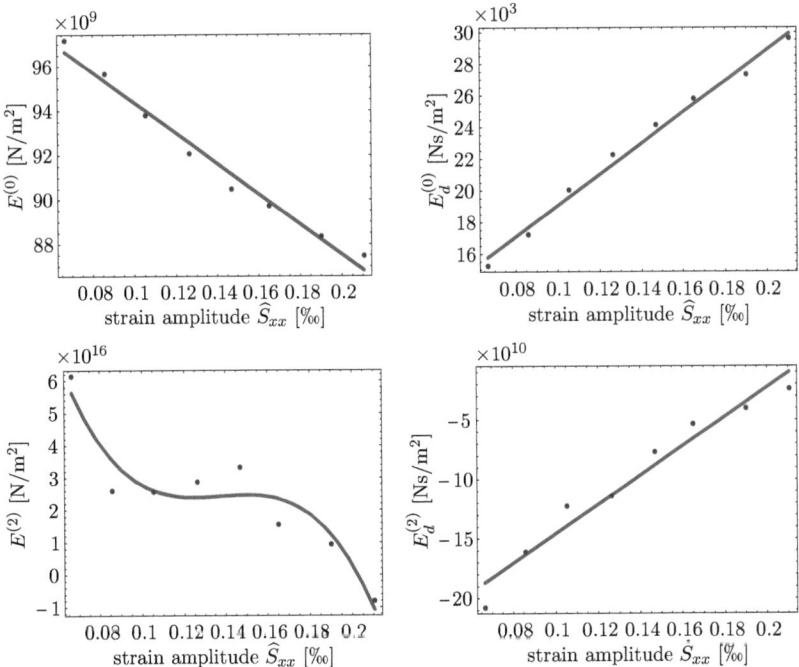

Figure 8.16: Linear and cubic mechanical parameters of PIC 181.

116   Chapter 8. Combination of nonlinear modelings

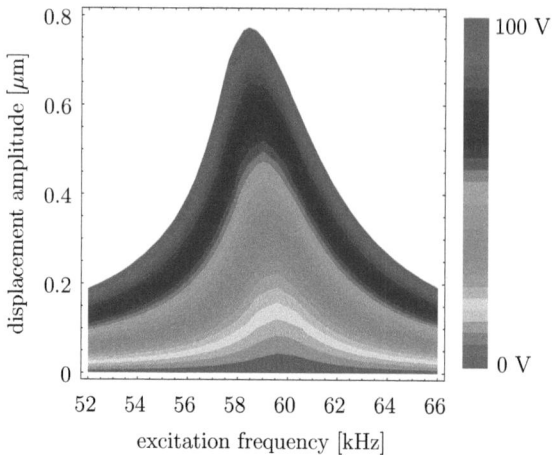

Figure 8.17: Amplitude response with cubic nonlinearities for PIC 181.

As a last example of the constitutive equation, the linear damping is omitted and a small change is made for the cubic dissipative term, we have

$$T_{xx} = E^{(0)} S_{xx} + E^{(2)} S_{xx}^3 + E_d^{(2)} \dot{S}_{xx}^3. \tag{8.23}$$

It can be seen in figure 8.18 that the theoretical stress–strain curves are now acute at both ends but an acceptable accordance of them with the experimental results holds. The identified parameters with the corresponding fitted curves are shown in figure 8.19, where a linear function is applied for $E^{(0)}$, a cubic function for $E^{(2)}$ and a quadratic function for $E_d^{(2)}$. Figure 8.20 shows the respective displacement amplitude responses.

In order to verify the above-calculated results, the experimental displacement amplitude response of a piezoceramic rod of PIC 181 excited near to resonance at 40 V is once again plotted in figure 8.21. It can be seen that for all cases of the constitutive equation the first resonance frequency is shifted from an experimental value 55.8 kHz to that of the modeling results 60 kHz approximately. This stems possibly from the academic establishment of the stress–strain hysteresis relations. If it is possible to overcome the difficulty of the tension tests, fitting the original measuring stress–strain loops may result in the elastic modulus of piezoceramics which is smaller in value than that so far, then the suitable resonance frequency can be derived. Due to the presence of high linear damping the jump phenomena is missing, and the resonance amplitude is only about 20% of the experimental value. From figures 8.11, 8.13 and 8.17 it is evident that the linear damping is dominant over the dissipative nonlinear terms. So far the best modeling result obtained from the last constitutive equation shown in figure 8.20 is good not only with regard to quality, revealing typical dynamic nonlinearities such as jump phenomenon and multiple stable responses, but also basically owning quantitatively an appropriate order of magnitude of the displacement amplitude.

Chapter 8. Combination of nonlinear modelings 117

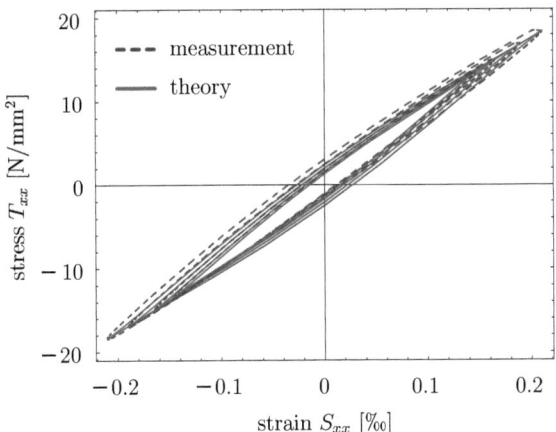

Figure 8.18: Comparision of the stress–strain hysteresis loops for PIC 181.

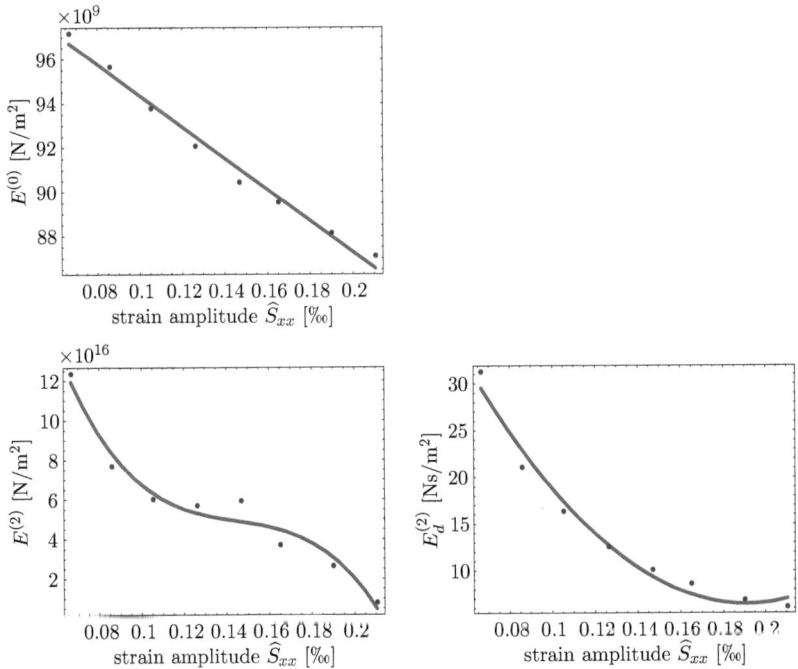

Figure 8.19: Elastic modulus and cubic mechanical parameters of PIC 181.

118   Chapter 8. Combination of nonlinear modelings

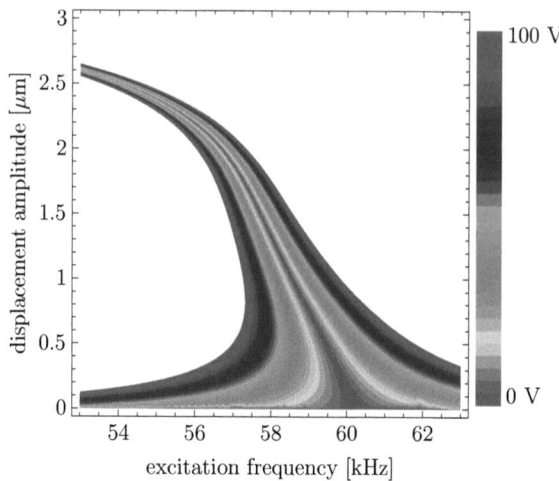

Figure 8.20: Amplitude response with cubic nonlinearities but without linear damping.

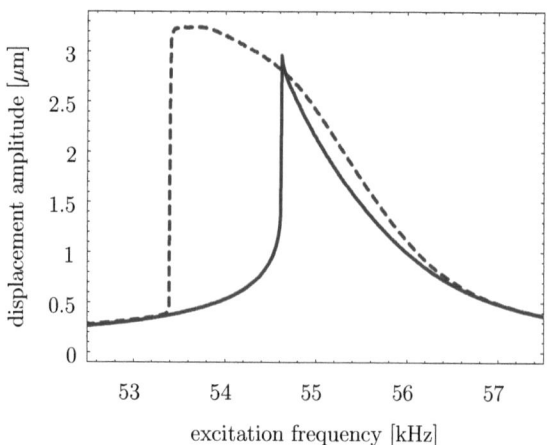

Figure 8.21: Experimental amplitude response for PIC 181 at 40 V. Solid line: sweep up of the excitation frequency, dashed line: sweep down.

Chapter 9

Conclusion and outlook

Following the results of von Wagner [117] this work intends to find consistent description of nonlinear dynamic behavior of piezoceramic actuators in resonance operation under weak electric fields far away from coercive ones. In chapter 2, nonlinear effects occurring in piezoceramic PZT materials subjected to strong quasi-static electrical or mechanical loads has been described referring to the results of Kamlah [63]. In this case the nonlinear behavior of piezoceramics results from polarization switching processes and reveals itself e.g. as dielectric or butterfly hysteresis when relations between applied electric fields and polarizations or strains are considered respectively.

By contrast, in chapter 5 nonlinear phenomena exhibited by piezoceramics in resonance operation are described, where moderate strains can be produced by weak applied electric fields. Typical dynamic nonlinearities, for instance jump phenomena in the stationary amplitude, multiple stable response at the same excitation as well as a decrease of the normalized response amplitude with increasing excitation amplitude can experimentally be observed as described in chapter 3. This nonlinear behavior is usually modeled by introducing higher-order conservative and dissipative terms into the piezoelectric constitutive equations, giving good qualitative and quantitative results in accordance with experiments.

Nevertheless, due to the ambiguity in the role of the type of nonlinearities and in the description of the damping, quasi-static experiments described in chapter 6 have been carried out with moderate electrical or mechanical loads respectively resulting in strains of the same order as those in the dynamic experiments. In both cases, piezoceramics exhibit hysteretic behavior that can well be modeled by using the classical Preisach model, the Prandtl-Ishlinskii model, the Masing model and the Bouc-Wen model.

Introducing the Masing model into the linear conservative dynamic modeling presented in chapter 4 leads only to qualitatively appropriate results. For the same use of the Bouc-Wen model, unsuitable results are derived. On the other hand, applying the nonlinear dynamic model with the mechanical parameters identified from the stress–strain hysteretic relations brings out good results in both quality and quantity. From the results of this work it can be concluded, that mechanical nonlinearities from the stress–strain relation play a dominant role in the explanation of the observed nonlinear dynamical behavior.

The nonlinear dynamic effects presented in this work correspond to the inverse 31-effect, so the above conclusion should be verified in the case when polarized piezoceramics excited using the inverse 33- or 15-effect are considered as described respectively in [89,119]. There the phenomena and the order of nonlinearities are absolutely similar to those described for the inverse 31-effect. With similar assumptions as mentioned in the present work, coupling and dielectric cubic nonlinear parameters (if existent) may also have an influence on the zeroth order approximation of displacement responses in the case of the inverse 33- and 15-effects. On the other hand, the systems under investigation are subjected to very weak external electric fields, owing to the resonance excitation and slight damping, giving rise to moderate strains and stresses, but both of them are still far away from the nonlinear effects described in section 2.1. Therefore it can also be expected for the case of the inverse 33- and 15-effects, that nonlinear stress–strain relations play a dominant role in the nonlinear dynamic effects. In addition, the piezoelectric coupling coefficient d_{15} of many piezoceramic materials, such as PIC 255 and PIC 181, is greater than d_{33} and d_{31} in absolute value, thus the inverse 15-effect will be useful to technical applications.

In chapter 7 the nonlinear hysteresis behavior of piezoceramics under moderate quasi-static electric fields has well been modeled. It is assumed that this nonlinearity does not influence the nonlinear dynamic effects since the electric fields in the dynamic case are still much smaller than the moderate fields (about 10%). However, in further work the combination of hysteresis effects in both pure mechanical and coupling aspects should be considered to describe the nonlinear dynamic behavior of piezoceramics.

In the study so far, due to the great difficulty in performing tension and compression tests, the modeling has not given accurate description of experimental results yet. Hence, the most necessity for further investigation is improving the precision of the corresponding tests. New clamping devices can be designed with respect to appropriate forms of samples, for instance samples with borings and/or in bone-shaped form. Longer samples can also be used to enhance the accuracy of measurements with laser extensometer. On the other hand, an extension to various materials, such as "hard" and "soft" PZT piezoceramics, lead-free or low-lead piezoceramics and to different use of piezoelectric effects can simultaneously be taken into account. From that answers for some remaining questions are expected to be found. For example, a few of such problems are given as following:

- Type of the nonlinearities.
- Quadratic or cubic nonlinearities.
- Modeling of the damping.
- Influence of the boundary conditions and the surroundings on the nonlinearities.
- Behavior of lead-free or low-lead piezoceramics.

Bibliography

[1] ALATSATHIANOS, S.: *Experimentelle Untersuchung des Materialverhaltens von piezoelektrischen Werkstoffen.* Ph.D. thesis, Universität Karlsruhe, 2000.

[2] ALTAY, G. A. AND DÖKMECI, M. C.: A non-linear rod theory for high-frequency vibrations of thermopiezoelectric materials. *International Journal of Non-Linear Mechanics*, 37 (2), 225–243, 2002.

[3] ANG, W. T.; KHOSLA, P. K. AND RIVIERE, C. N.: Feedforward Controller With Inverse Rate-Dependent Model for Piezoelectric Actuators in Trajectory-Tracking Applications. *IEEE/ASME Transactions on Mechatronics*, 12 (2), 134–142, 2007.

[4] ANTON, S. R. AND SODANO, H. A.: A review of power harvesting using piezoelectric materials (2003–2006). *Smart Materials and Structures*, 16 (3), R1–R21, 2007.

[5] ARAFA, M. AND BAZ, A.: On the Nonlinear Behavior of Piezoelectric Actuators. *Journal of Vibration and Control*, 10 (3), 387–398, 2004.

[6] AROCKIARAJAN, A.: *Computational Modeling of Domain Switching Effects in Piezoceramic Materials – A Micro-Macro Mechanical Approach.* Ph.D. thesis, Technische Universität Kaiserslautern, 2005.

[7] AROCKIARAJAN, A.; MENZEL, A.; DELIBAS, B. AND SEEMANN, W.: Micromechanical Modeling of Switching Effects in Piezoelectric Materials - A Robust Coupled Finite Element Approach. *Journal of Intelligent Material Systems and Structures*, 18 (9), 983–999, 2007.

[8] AROCKIARAJAN, A.; MENZEL, A. AND SEEMANN, W.: Constitutive modelling of rate-dependent domain switching effects in ferroelectric and ferroelastic materials. *Journal of Electroceramics*, 20 (3-4), 159–165, 2008.

[9] BALLAS, R. G.: *Piezoelectric Multilayer Beam Bending Actuators - Static and Dynamic Behavior and Aspects of Sensor Integration.* Springer-Verlag, Berlin, Heidelberg, 2007.

[10] BASHASH, S. AND JALILI, N.: Robust Adaptive Control of Coupled Parallel Piezo-Flexural Nanopositioning Stages. *IEEE/ASME Transactions on Mechatronics*, 14 (1), 11–20, 2009.

[11] BEIGE, H.: Elastic and dielectric nonlinearities of piezoelectric ceramics. *Ferroelectrics*, 51 (1), 113–119, 1983.

[12] BEIGE, H. AND SCHMIDT, G.: Electromechanicals resonances for investigating linear and nonlinear properties of dielectrics. *Ferroelectrics*, 41 (1), 39–49, 1982.

[13] BERTOTTI, G.: *Hysteresis in Magnetism*. Academic Press, 1998.

[14] BIORCI, G. AND PESCETTI, D.: Analytical Theory of the Behaviour of Ferromagnetic Materials. *Il Nuovo Cimento*, 7 (6), 829–842, 1958.

[15] BIORCI, G. AND PESCETTI, D.: Some consequences of the analytical theory of the ferromagnetic hysteresis. *Journal de Physique et le Radium*, 20 (2-3), 233–236, 1959.

[16] BIORCI, G. AND PESCETTI, D.: Some Remarks on Hysteresis. *Journal of Applied Physics*, 37, 425–427, 1966.

[17] BOUC, R.: Forced vibrations of a mechanical system with hysteresis. In *Proceedings of the 4th Conference on Nonlinear Oscillations*, p. 315. Prague, Czechoslovakia, 1967. (Abstract).

[18] BOUC, R.: Modèle mathématique d'hystérésis. *Acustica*, 24, 16–25, 1971.

[19] CADY, W. G.: *Piezoelectricity : An introduction to the Theory and Applications of Electromechanical Phenomena in Crystals*. McGraw-Hill, NewYork, 1946.

[20] CERAVOLO, R.; DEMARIE, G. V. AND ERLICHER, S.: Instantaneous identification of Bouc-Wen-type hysteretic systems from seismic response data. *Key Engineering Materials*, 347 (Damage Assessment of Structures VII), 331–338, 2007.

[21] CHEN, X. AND HISAYAMA, T.: Adaptive Sliding-Mode Position Control for Piezo-Actuated Stage. *IEEE Transactions on Industrial Electronics*, 55 (11), 3927–3934, 2008.

[22] CLAEYSSEN, F.; JÄNKER, P.; LELETTY, R.; SOSNIKI, O.; PAGES, A.; MAGNAC, G. AND CHRISTMANN, M.: New Actuators for Aircraft, Space and Military Applications. In *Proceedings of ACTUATOR 2010, 12th International Conference on New Actuators*, pp. 324–330. Bremen, Germany, 2010.

[23] CURIE, J. AND CURIE, P.: Développement, par pression, de l'électricité polaire dans les cristaux hémièdres à faces inclinées. *Comptes rendus de l'Académie des sciences*, 91, 294–295, 1880.

[24] DE VOLDER, M. AND REYNAERTS, D.: Pneumatic and hydraulic microactuators: a review. *Journal of Micromechanics and Microengineering*, 20 (4), 18 pp., 2010.

[25] DELIBAS, B.: *Rate dependant nonlinear properties of perovskite tetragonal piezoelectric materials using a micromechanical model*. Ph.D. thesis, Technische Universität Kaiserslautern, 2005.

[26] EDERY-AZULAY, L. AND ABRAMOVICH, H.: The integrity of piezo-composite beams under high cyclic electro-mechanical loads–experimental results. *Smart Materials and Structures*, 16 (4), 1226–1238, 2007.

[27] EVERETT, D. H.: A general approach to hysteresis. Part 3.-A formal treatment of the independent domain model of hysteresis. *Transactions of the Faraday Society*, 50, 1077–1096, 1954.

[28] EVERETT, D. H.: A general approach to hysteresis. Part 4. An alternative formulation of the domain model. *Transactions of the Faraday Society*, 51, 1551–1557, 1955.

[29] EVERETT, D. H. AND SMITH, F. W.: A general approach to hysteresis. Part 2: Development of the domain theory. *Transactions of the Faraday Society*, 50, 187–197, 1954.

[30] EVERETT, D. H. AND WHITTON, W. I.: A general appoach to hysteresis. *Transactions of the Faraday Society*, 48, 749–757, 1952.

[31] FETT, T.; MÜLLER, S.; MUNZ, D. AND THUN, G.: Nonsymmetry in the deformation behaviour of PZT. *Journal of Materials Science Letters*, 17 (4), 261–265, 1998.

[32] FETT, T.; MUNZ, D. AND THUN, G.: Nonsymmetric Deformation Behavior of Lead Zirconate Titanate Determined in Bending Tests. *Journal of the American Ceramic Society*, 81 (1), 269–272, 1998.

[33] FETT, T.; MUNZ, D. AND THUN, G.: Nonsymmetric deformation behavior of several PZT ceramics. *Journal of Materials Science Letters*, 18 (20), 1641–1643, 1999.

[34] FETT, T.; MUNZ, D. AND THUN, G.: Polarization measurements on PZT under transverse tensile loading. *Ferroelectrics*, 247 (4), 321–332, 2000.

[35] GAUL, L. AND BECKER, J.: Model-based piezoelectric hysteresis and creep compensation for highly-dynamic feedforward rest-to-rest motion control of piezoelectrically actuated flexible structures. *International Journal of Engineering Science*, 47 (11-12), 1193–1207, 2009.

[36] GE, P. AND JOUANEH, M.: Modeling hysteresis in piezoceramic actuators. *Precision Engineering*, 17 (3), 211–221, 1995.

[37] GOLDFARB, M. AND CELANOVIC, N.: Modeling Piezoelectric Stack Actuators for Control of Micromanipulation. *IEEE Control Systems*, 17 (3), 69–79, 1997.

[38] GOMIS-BELLMUNT, O.; IKHOUANE, F. AND MONTESINOS-MIRACLE, D.: Control of Bouc-Wen hysteretic systems: Application to a piezoelectric actuator. In *Proceedings of the 13th EPE-PEMC 2008, Power Electronics and Motion Control Conference*, pp. 1670–1675. Poznan, 2008.

[39] GORBET, R. B.; WANG, D. W. L. AND MORRIS, K. A.: Preisach Model Identification of a Two-Wire SMA Actuator. In *Proceedings of IEEE International Conference on Robotics and Automation*, pp. 2161–2167. Leuven, Belgium, May 1998.

[40] HAGEDORN, P.: *Nichtlineare Schwingungen*. Akademische Verlagsgesellschaft, Wiesbaden, 1978.

[41] HAGEDORN, P.: *Technische Schwingungslehre Band 2: Linear Schwingungen kontinuierlicher mechanischer Systeme*. Springer-Verlag, Berlin, Heidelberg, New York, London, Paris, Tokyo, 1978.

[42] HAGEDORN, P. AND WALLASCHEK, J.: Travelling Wave Ultrasonic Motors, Part I: Working Principle and Mathematical Modelling of the Stator. *Journal of Sound and Vibration*, 155 (1), 31–46, 1992.

[43] HAGOOD, N. W. AND HORODEZKY, J.: Method and apparatus for active control of golf club impact. *United States Patent No. 7,780,535 B2*, 2010.

[44] HEGEWALD, T.: *Modellierung des nichtlinearen Verhaltens piezokeramischer Aktoren*. Ph.D. thesis, Universität Erlangen-Nürnberg, 2007.

[45] HEGEWALD, T.; KALTENBACHER, B.; KALTENBACHER, M. AND LERCH, L.: Efficient Modeling of Ferroelectric Behavior for the Analysis of Piezoceramic Actuators. *Journal of Intelligent Material Systems and Structures*, 19 (10), 1117–1129, 2008.

[46] HENZE, O. AND RUCKER, W. M.: Identification procedures of Preisach model. *IEEE Transactions on Magnetics*, 38 (2), 833–836, March 2002.

[47] HEYWANG, W.; LUBITZ, K. AND WERSING, W. (Eds.): *Piezoelectricity*. Springer-Verlag, Berlin, Heidelberg, 2008.

[48] HOFFMANN, K.: *Eine Einführung in die Technik des Messens mit Dehnungsmessstreifen*. Hottinger Baldwin Messtechnik GmbH, Darmstadt, 1987.

[49] HOFFMANN, K.-H. AND MEYER, G. H.: A Least Squares Method for Finding the Preisach Hysteresis Operator from Measurements. *Numerische Mathematik*, 55, 695–710, 1989.

[50] HOFFMANN, K.-H.; SPREKELS, J. AND VISINTIN, A.: Identification of Hysteresis Loops. *Journal of Computational Physics*, 78 (1), 215–230, 1988.

[51] HWANG, S. C.; LYNCH, C. S. AND MCMEEKING, R. M.: Ferroelectric/ferroelastic interactions and a polarization switching model. *Acta Metallurgica et Materialia*, 43 (5), 2073–2084, 1995.

[52] IKEDA, T.: *Fundamentals of Piezoelectricity*. Oxford University Press, Oxford, New York, Tokyo, 1990.

[53] ISMAIL, M.; IKHOUANE, F. AND RODELLAR, J.: The Hysteresis Bouc-Wen Model, a Survey. *Archives of Computational Methods in Engineering*, 16, 161–188, 2009.

[54] IYER, R. V. AND SHIRLEY, M. E.: Hysteresis Parameter Identification With Limited Experimental Data. *IEEE Transactions on Magnetics*, 40 (5), 3227–3239, 2004.

[55] JAFFE, B.; COOK, W. R. AND JAFFE, H.: *Piezoelectric ceramics*. Academic Press, London, New York, 1971.

[56] JANOCHA, H. (Ed.): *Adaptronics and Smart Structures - Basics, Materials, Design and Applications*. Springer-Verlag, Berlin, Heidelberg, New York, 2007.

[57] JANOCHA, H. AND KUHNEN, K.: Real-time compensation of hysteresis and creep in piezoelectric actuators. *Sensors and Actuators A: Physical*, 79 (2), 83–89, 2000.

[58] JAYAKUMAR, K.; YADAV, D. AND NAGESWARA RAO, B.: Nonlinear vibration analysis for a generic coupled-laminated plate with surface bonded or embedded induced strain actuators. *Journal of Sound and Vibration*, 301 (3-5), 846–863, 2007.

[59] JAYAKUMAR, K.; YADAV, D. AND NAGESWARA RAO, B.: Nonlinear free vibration analysis of generic coupled induced strain actuated piezo-laminated beams. *Forschung im Ingenieurwesen*, 72 (3), 153–162, 2008.

[60] JAYAKUMAR, K.; YADAV, D. AND NAGESWARA RAO, B.: Nonlinear free vibration analysis of simply supported piezo-laminated plates with random actuation electric potential difference and material properties. *Communications in Nonlinear Science and Numerical Simulation*, 14 (4), 1646–1663, 2009.

[61] JOSHI, S. P.: Non-linear constitutive relations for piezoceramic materials. *Journal of Smart Materials and Structures*, 1 (1), 80–83, 1992.

[62] JUNG, J. AND HUH, K.: Simulation tool design for the two-axis nano stage of lithography systems. *Mechatronics*, 20 (5), 574–581, 2010.

[63] KAMLAH, M.: Ferroelectric and ferroelastic piezoceramics - modeling of electromechanical hysteresis phenomena. *Continuum Mech. Thermodyn.*, 13, 219–268, 2001.

[64] KAMLAH, M. AND JIANG, Q.: A constitutive model for ferroelectric PZT ceramics under uniaxial loading. *Smart Materials and Structures*, 8 (4), 441–459, 1999.

[65] KOLSCH, H.: *Schwingungsdämpfung durch statische Hysterese - Modellierung von Bauteilen, Parameteridentifikation, Schwingungsberechnungen*. Fortschr.-Ber. VDI Reihe 11 Nr. 190, VDI-Verlag, Düsseldorf, 1993.

[66] KTENA, A.; FOTIADIS, D. I.; SPANOS, P. D. AND MASSALAS, C. V.: A Preisach model identification procedure and simulation of hysteresis in ferromagnets and shape-memory alloys. *Physica B: Condensed Matter*, 306 (1-4), 84–90, 2001.

[67] KUHNEN, K. AND KREJCI, P.: Compensation of Complex Hysteresis and Creep Effects in Piezoelectrically Actuated Systems - A New Preisach Modeling Approach. *IEEE Transactions on Automatic Control*, 54 (3), 537–550, 2009.

[68] LEE, S.-H.; OZER, M. B. AND ROYSTON, T. J.: Hysteresis Models for Piezoceramic Transducers. *Journal of Materials Processing and Manufacturing Science*, 9, 33–52, July 2000.

[69] LEE, S.-H.; OZER, M. B. AND ROYSTON, T. J.: Piezoceramic Hysteresis in the Adaptive Structural Vibration Control Problem. *Journal of Intelligent Material Systems and Structures*, 13, 117–124, February/March 2002.

[70] LEE, S.-H. AND ROYSTON, T. J.: Modeling piezoceramic transducer hysteresis in the structural vibration control problem. *Journal of Acoustical Society of America*, 108 (6), 2843–2855, December 2000.

[71] LEE, S.-H.; ROYSTON, T. J. AND FRIEDMAN, G.: Modeling and Compensation of Hysteresis in Piezoceramic Transducers for Vibration Control. *Journal of Intelligent Material Systems and Structures*, 11, 781–790, October 2000.

[72] LEE, S. K.; KIM, Y. S.; PARK, H. C.; YOON, K. J.; GOO, N. S.; YU, Y. AND CHO, C.: Performance analysis of a lightweight piezo-composite actuator considering the material non-linearity of an embedded PZT wafer. *Smart Materials and Structures*, 14 (6), 1101–1106, 2005.

[73] LI, S.; SUN, S.-J.; LIU, D.-C.; LIN, S.-P.; JUANG, D.-P.; WANG, C.-H. AND GER, G.-S.: Vibration suppressed bicycle structure. *United States Patent No. 6,986,521 B1*, 2006.

[74] LI, Y. AND XU, Q.: Adaptive Sliding Mode Control With Perturbation Estimation and PID Sliding Surface for Motion Tracking of a Piezo-Driven Micromanipulator. *IEEE Transactions on Control Systems Technology*, 18 (4), 798–810, 2010.

[75] LIN, C. J. AND CHEN, S. Y.: Evolutionary Algorithm Based Feedforward Control for Contouring of a Biaxial Piezo-Actuated Stage. *NEMS/MEMS Technology and Devices, Advanced Materials Research*, 74, 235–238, 2009.

[76] LIPPMANN, M. G.: Principe de la conservation de l'électricité. *Annales de chimie et de physique*, 24 (5), 145–178, 1881.

[77] LUPASCU, D. C. AND RÖDEL, J.: Fatigue in Bulk Lead Zirconate Titanate Actuator Materials. *Advanced Engineering Materials*, 7 (10), 882–898, 2005.

[78] MASING, G.: Zur Heynschen Theorie der Verfestigung der Metalle durch verborgene elastische Spannungen. *Wissenschaftliche Veröffentlichungen aus dem Siemens-Konzern*, 3 (1), 231–239, 1923/24.

[79] MAUGIN, G. A.: *Nonlinear Electromechanical Effects and Applications*. World Scientific, 1985.

[80] MAUGIN, G. A.: *Continuum Mechanics of Electromagnetic Solids*. Elsevier Science Publishing, 1988.

[81] MAYERGOYZ, I. D.: *Mathematical Models of Hysteresis*. Springer-Verlag, New York, Berlin, Heidelberg, London, Paris, Tokyo, Hong Kong, Barcelona, 1991.

[82] MEEKER, T. R. (Ed.): *ANSI/IEEE Std 176-1987, An American National Standard, IEEE Standard on Piezoelectricity.* The Institute of Electrical and Electronic Engineers, Inc., New York, 1988.

[83] MEHLFELDT, D.: *Modellierung und optimale Steuerung piezoelektrisch aktuierter Einspritzventile.* Ph.D. thesis, Universität Siegen, 2006.

[84] MEHLING, V.; TSAKMAKIS, C. AND GROSS, D.: Phenomenological model for the macroscopical material behaviour of ferroelectric ceramics. *Journal of the Mechanics and Physics of Solids*, 55 (10), 2106–2141, 2007.

[85] MERRY, R.; UYANIK, M.; VAN DE MOLENGRAFT, R.; KOOPS, R.; VAN VEGHEL, M. AND STEINBUCH, M.: Identification, Control and Hysteresis Compensation of a 3 DOF Metrological AFM. *Asian Journal of Control*, 11 (2), 130–143, 2009.

[86] MOHEIMANI, S. O. R. AND FLEMING, A. J.: *Piezoelectric Transducers for Vibration Control and Damping.* Springer-Verlag, London, 2006.

[87] NAYFEH, A. H. AND MOOK, D.: *Nonlinear Oscillatons.* John Wiley & Sons, New York, Chichester, Brisbane, Toronto, Singapore, 1979.

[88] OTTL, D.: *Schwingungen mechanischer Systeme mit Strukturdämpfung.* VDI-Forschungsheft Nr. 603, VDI-Verlag, Düsseldorf, 1981.

[89] PARASHAR, S. K.; DASGUPTA, A.; VON WAGNER, U. AND HAGEDORN, P.: Non-linear shear vibrations of piezoceramic actuators. *International Journal of Nonlinear Mechanics*, 40 (4), 429–443, 2005.

[90] PARASHAR, S. K. AND VON WAGNER, U.: Nonlinear Longitudinal Vibrations of Transversally Polarized Piezoceramics: Experiments and Modeling. *Nonlinear Dynamics*, 37, 51–73, 2004.

[91] PARASHAR, S. K.; VON WAGNER, U. AND HAGEDORN, P.: A Modified Timoshenko Beam Theory for Nonlinear Shear-Induced Flexural Vibrations of Piezoceramic Continua. *Nonlinear Dynamics*, 37, 181–205, 2004.

[92] PARASHAR, S. K.; VON WAGNER, U. AND HAGEDORN, P.: Nonlinear shear-induced flexural vibrations of piezoceramic actuators: experiments and modeling. *Journal of Sound and Vibration*, 285 (4-5), 989–1014, 2005.

[93] PASCO, Y. AND BERRY, A.: A Hybrid Analytical/Numerical Model of Piezoelectric Stack Actuators Using a Macroscopic Nonlinear Theory of Ferroelectricity and a Preisach Model of Hysteresis. *Journal of Intelligent Material Systems and Structures*, 15 (5), 375–386, 2004.

[94] PESOTSKI, D.; JANOCHA, H. AND KUHNEN, K.: Adaptive Compensation of Hysteretic and Creep Non-linearities in Solid-state Actuators. *Journal of Intelligent Material Systems and Structures*, 21 (14), 1437–1446, 2010.

[95] PREISACH, F.: Über die magnetische Nachwirkung. *Zeitschrift für Physik*, 94, 277–302, 1935.

[96] PRUKSANUBAL, P.; BINNER, A. AND GONSCHOREK, K. H.: Determination of distribution functions and parameters for the Preisach hysteresis model. In *Proceedings of 17th International Zurich Symposium on Electromagnetic Compatibility*, pp. 258–261. EMC-Zurich, 2006.

[97] RECHDAOUI, M. S. AND AZRAR, L.: Active control of secondary resonances piezoelectric sandwich beams. *Applied Mathematics and Computation*, 216 (11), 3283–3302, 2010.

[98] RECHDAOUI, M. S.; AZRAR, L.; BELOUETTAR, S.; DAYA, E. M. AND POTIER-FERRY, M.: Active Vibration Control of Piezoelectric Sandwich Beams at Large Amplitudes. *Mechanics of Advanced Materials and Structures*, 16 (2), 98–109, 2009.

[99] ROGACHEVA, N. N.: *The theory of piezoelectric shells and plates*. CRC Press, Boca Raton, Ann Arbor, London, Tokyo, 1994.

[100] ROYSTON, T. J.; LEE, S.-H. AND FRIEDMAN, G.: Comparision of two rate-independent hysteresis models with application to piezoceramic transducers. In *Proceedings of ASME Design Engineering Technical Conference Symposium on Nonlinear Response of Hysteretic Oscillators*. Las Vegas, NV, September 1999.

[101] SATEESH, V. L.; UPADHYAY, C. S. AND VENKATESAN, C.: Thermodynamic Modeling of Hysteresis Effects in Piezoceramics for Application to Smart Structures. *AIAA Journal*, 46 (1), 280–284, 2008.

[102] SATEESH, V. L.; UPADHYAY, C. S. AND VENKATESAN, C.: Nonlinear Analysis of Smart Composite Plates Including Hysteresis Effects. *AIAA Journal*, 48 (9), 2017–2028, 2010.

[103] SATEESH, V. L.; UPADHYAY, C. S. AND VENKATESAN, C.: A study of the polarization–electric-field nonlinear effect on the response of smart composite plates. *Smart Materials and Structures*, 19 (7), 1–16, 2010.

[104] SATTEL, T.: *Dynamics of Ultrasonic Motors*. Ph.D. thesis, Technische Universität Darmstadt, 2002.

[105] SAUSER, B.: Energy-Saving Helicopter Blades: NASA researchers are using smart materials to improve helicopter performance. *Technology Review (MIT)*, http://www.technologyreview.com/computing/22278/?a=f, March 12, 2009.

[106] SHAN, Y. AND LEANG, K. K.: Repetitive Control with Prandtl-Ishlinskii Hysteresis Inverse for Piezo-Based Nanopositioning. In *Proceedings of the 2009 American Control Conference*, pp. 301–306. St. Louis, MO, USA, 2009.

[107] SHIRLEY, M. E.: *Identification of Preisach Measures*. Master's thesis, Texas Tech University, 2003.

[108] STANTON, S. C.; ERTURK, A.; MANN, B. P. AND INMAN, D. J.: Nonlinear piezoelectricity in electroelastic energy harvesters: Modeling and experimental identification. *Journal of Applied Physics*, 108 (074903), 9 pp., 2010.

[109] TAN, P. AND TONG, L.: A one-dimensional model for non-linear behaviour of piezoelectric composite materials. *Composite Structures*, 58 (4), 551–561, 2002.

[110] TAN, X.: *Control of Smart Actuators*. Ph.D. thesis, University of Maryland, 2002.

[111] TIERSTEN, H. F.: *Linear Piezoelectric Plate Vibrations*. Plenum Press, New York, 1969.

[112] UCHINO, K.: Piezoelectric actuators 2006 – Expansion from IT/robotics to ecological/energy applications. *Journal of Electroceramics*, 20 (3-4), 301–311, 2008.

[113] USHER, T. AND SIM, A.: Nonlinear dynamics of piezoelectric high displacement actuators in cantilever mode. *Journal of Applied Physics*, 98 (064102), 2005.

[114] VANDERGRIFT, J. A.: Piezoelectric damper for a board such as a snow ski or snowboard. *United States Patent No. 5,775,715*, 1998.

[115] VANDERGRIFT, J. A.; DEROCCO, A. O.; GLENNE, B. AND STERLING, S. M. K.: Active piezoelectric damper for a snow ski or snowboard. *United States Patent No. 6,095,547*, 2000.

[116] VISINTIN, A.: *Differential Models of Hysteresis*. Springer-Verlag, Berlin, Heidelberg, 1994.

[117] VON WAGNER, U.: *Nichtlineare Effekte bei Piezokeramiken unter schwachem elektrischem Feld: Experimentelle Untersuchung und Modellbildung*. Habilitationsschrift TU Darmstadt, GCA-Verlag, Herdecke, 2003.

[118] VON WAGNER, U.: Non-linear longitudinal vibrations of piezoceramics excited by weak electric fields. *International Journal of Non-Linear Mechanics*, 38 (4), 565–574, 2003.

[119] VON WAGNER, U.: Non-linear longitudinal vibrations of non-slender piezoceramic rods. *International Journal of Non-Linear Mechanics*, 39 (4), 673–688, 2004.

[120] VON WAGNER, U. AND HAGEDORN, P.: Piezo-Beam Systems Subjected to Weak Electric Field: Experiments and Modeling of Non-linearities. *Journal of Sound and Vibration*, 256 (5), 861–872, 2002.

[121] VON WAGNER, U. AND HAGEDORN, P.: Nonlinear Effects of Piezoceramics Excited by Weak Electric Fields. *Nonlinear Dynamics*, 31, 133–149, 2003.

[122] VON WAGNER, U.; HAGEDORN, P. AND NGUYEN, M. N.: Nonlinear Behavior of Piezo-Beam-Systems Subjected to Weak Electric Field. In *Proceedings of DETC2001, ASME Design Engineering Technical Conference*, VIB21488. Pittsburgh, 2001.

[123] WANG, Z. Y. AND MAO, J. Q.: On PSO based Bouc-Wen modeling for piezoelectric actuator. In *Proceedings of ICIRA'10 the Third international conference on Intelligent robotics and applications - Volume Part I*, pp. 125–134. Shanghai, China, 2010.

[124] WEBER, M.-A.; KAMLAH, M. AND MUNZ, D.: *Experimente zum Zeitverhalten von Piezokeramiken.* Forschungszentrum Karlsruhe Technik und Umwelt, Wissenschaftliche Berichte FZKA 6465, Karlsruhe, 2000.

[125] WEN, Y.-K.: Method for random vibration of hysteretic systems. *ASCE Journal of the Engineering Mechanics Division*, 102, 249–263, 1976.

[126] WOLF, F.; SUTOR, A.; RUPITSCH, S. J. AND LERCH, R.: Modeling and measurement of hysteresis of ferroelectric actuators considering time-dependent behavior. *Procedia Engineering (Proc. Eurosensor XXIV Conference)*, 5, 87–90, 2010.

[127] WOLF, K.: *Electromechanical Energy Conversion in Asymmetric Piezoelectric Bending Actuators.* Ph.D. thesis, Technische Universität Darmstadt, 2000.

[128] WOLF, K. AND GOTTLIEB, O.: Nonlinear dynamics of a noncontacting atomic force microscope cantilever actuated by a piezoelectric layer. *Journal of Applied Physics*, 91 (7), 4701–4709, 2002.

[129] WU, Y. AND ZOU, Q.: Iterative Control Approach to Compensate for Both the Hysteresis and the Dynamics Effects of Piezo Actuators. *IEEE Transactions on Control Systems Technology*, 15 (5), 936–944, 2007.

[130] YEH, T.-J.; LU, S.-W. AND WU, T.-Y.: Modeling and Identification of Hysteresis in Piezoelectric Actuators. *Journal of Dynamic Systems, Measurement, and Control*, 128 (2), 189–196, 2006.

[131] ZHOU, D.: *Experimental Investigation of Non-linear Constitutive Behavior of PZT Piezoceramics.* Ph.D. thesis, Universität Karlsruhe (TH), 2003.

[132] ZHOU, D.; KAMLAH, M. AND MUNZ, D.: Effects of Bias Electric Fields on the Non-linear Ferroelastic Behavior of Soft Lead Zirconate Titanate Piezoceramics. *Journal of the American Ceramic Society*, 88 (4), 867–874, 2005.

[133] ZHOU, X. AND CHATTOPADHYAY, A.: Hysteresis Behavior and Modeling of Piezoceramic Actuators. *Journal of Applied Mechanics*, 68, 270–277, 2001.

Die VDM Verlagsservicegesellschaft sucht für wissenschaftliche Verlage abgeschlossene und herausragende

Dissertationen, Habilitationen, Diplomarbeiten, Master Theses, Magisterarbeiten usw.

für die kostenlose Publikation als Fachbuch.

Sie verfügen über eine Arbeit, die hohen inhaltlichen und formalen Ansprüchen genügt, und haben Interesse an einer honorarvergüteten Publikation?

Dann senden Sie bitte erste Informationen über sich und Ihre Arbeit per Email an *info@vdm-vsg.de*.

Sie erhalten kurzfristig unser Feedback!

VDM Verlagsservicegesellschaft mbH
Dudweiler Landstr. 99　　　　　Telefon +49 681 3720 174
D - 66123 Saarbrücken　　　　　Fax　　　+49 681 3720 1749
www.vdm-vsg.de

Die VDM Verlagsservicegesellschaft mbH vertritt

Printed by Books on Demand GmbH, Norderstedt / Germany